# Plastics and Microplastics

Recent Titles in the
# CONTEMPORARY WORLD ISSUES
Series

*Racism in America: A Reference Handbook*
Steven L. Foy

*Waste Management: A Reference Handbook*
David E. Newton

*Sexual Harassment: A Reference Handbook*
Merril D. Smith

*The Climate Change Debate: A Reference Handbook*
David E. Newton

*Voting Rights in America: A Reference Handbook*
Richard A. Glenn and Kyle L. Kreider

*Modern Slavery: A Reference Handbook*
Christina G. Villegas

*Race and Sports: A Reference Handbook*
Rachel Laws Myers

*World Oceans: A Reference Handbook*
David E. Newton

*First Amendment Freedoms: A Reference Handbook*
Michael C. LeMay

*Medicare and Medicaid: A Reference Handbook*
Greg M. Shaw

*Organic Food and Farming: A Reference Handbook*
Shauna M. McIntyre

*Civil Rights and Civil Liberties in America: A Reference Handbook*
Michael C. LeMay

*GMO Food: A Reference Handbook, Second Edition*
David E. Newton

*Pregnancy and Birth: A Reference Handbook*
Keisha L. Goode and Barbara Katz Rothman

*Hate Groups: A Reference Handbook*
David E. Newton

Books in the **Contemporary World Issues** series address vital issues in today's society such as genetic engineering, pollution, and biodiversity. Written by professional writers, scholars, and nonacademic experts, these books are authoritative, clearly written, up-to-date, and objective. They provide a good starting point for research by high school and college students, scholars, and general readers as well as by legislators, businesspeople, activists, and others.

Each book, carefully organized and easy to use, contains an overview of the subject, a detailed chronology, biographical sketches, facts and data and/or documents and other primary source material, a forum of authoritative perspective essays, annotated lists of print and nonprint resources, and an index.

Readers of books in the Contemporary World Issues series will find the information they need in order to have a better understanding of the social, political, environmental, and economic issues facing the world today.

# Plastics and Microplastics

## A REFERENCE HANDBOOK

David E. Newton

An Imprint of ABC-CLIO, LLC

Santa Barbara, California • Denver, Colorado

Copyright © 2021 by ABC-CLIO, LLC

**Library of Congress Cataloging-in-Publication Data**

Names: Newton, David E., author.

Title: Plastics and microplastics : a reference handbook / David E. Newton.

Description: Santa Barbara, California : ABC-CLIO, [2021] | Series: Contemporary world issues | Includes bibliographical references and index.

Identifiers: LCCN 2021008696 (print) | LCCN 2021008697 (ebook) | ISBN 9781440875397 (hardcover) | ISBN 9781440875403 (ebook)

Subjects: LCSH: Plastics. | Microplastics.

Classification: LCC TP1120 .N484 2021 (print) | LCC TP1120 (ebook) | DDC 668.4—dc23

LC record available at https://lccn.loc.gov/2021008696

LC ebook record available at https://lccn.loc.gov/2021008697

ISBN: 978-1-4408-7539-7 (print)
      978-1-4408-7540-3 (ebook)

25 24 23 22 21     1 2 3 4 5

This book is also available as an eBook.

ABC-CLIO
An Imprint of ABC-CLIO, LLC

ABC-CLIO, LLC
147 Castilian Drive
Santa Barbara, California 93117
www.abc-clio.com

This book is printed on acid-free paper ∞

Manufactured in the United States of America

The United States produces about 234 pounds of plastic waste per person per year, second only to Germany in that category. Americans make up about 4 percent of the world population but produce more than 12 percent of the world's plastic wastes. They recycle about 9 percent of the plastic wastes they generate and incinerate another 16 percent. The remaining 75 percent of their plastic wastes end up in dumps and landfills. A large fraction of that trash eventually works its way into rivers, streams, lakes, and, eventually, the world's oceans.

Most people are aware of plastic pollution in the oceans. Newspapers and online websites frequently show pictures of whales and dolphins that have died because of the huge amounts of plastics they have ingested. Or they show disturbing pictures of turtles, seabirds, and other marine organisms that have become entangled in lost fishing nets, traps, and other fishing equipment, left to die because they are unable to eat or move about in their environment.

More recently, stories have begun to appear also about the dangers posed by microplastics, tiny pieces of plastic produced by the breakdown of plastic bags, clothing, tires, cups, eating utensils, building materials, and other substances made of plastic. Although research has just begun on this issue, many experts are already convinced that microplastics are likely to be toxic not only to marine and terrestrial organisms but also to humans.

This information has led to a growing interest in ways of dealing with plastics. Some efforts have been expended on developing very large systems by which wastes can be physically collected in the oceans and transported to sites on land. Other programs have been, and are being, developed to significantly increase the amount of plastic recycled and reused. More and more experts are beginning to talk about large-scale, revolutionary methods for dealing with all types of solid wastes, especially plastic wastes. For example, the zero waste movement has as its primary goal the complete discontinuation of the buildup of plastic wastes at all. A similar concept, the circular, or closed-loop, economy is a method for reducing the harmful effects of plastics at every stage of their life cycle, from extraction to transportation to production to use to reuse.

The challenges posed by plastic wastes are, however, overwhelming. In the first place, plastics are one of the most valuable materials ever discovered or invented by humans. They have a host of chemical, physical, electrical, and other properties that make them the substance of choice for a host of personal, commercial, and industrial products and operations. It is difficult to imagine exactly how we go about weaning ourselves away from some or many of the everyday applications in which plastics occur. Other forces drive the seemingly ever-increasing demand for plastics. For example, companies that have traditionally been based on the recovery and processing of fossil fuels are seeing growing consumer reluctance to continue on this line of business. The threats of climate change are one important element in the call for a reduction in the extraction and use of petroleum and natural gas. For these companies, the conversion of their operations to the production of plastics is an obvious, straightforward, and relatively simple process. So, while concerns about plastic waste grow, the forces for increased production also become more powerful.

This book is designed as a resource for a better understanding of this issue. Chapters 1 and 2 deal with the factual information needed for a better understanding of most aspects of the issue: the development of the plastic industry, options that have been developed for dealing with used plastics, threats posed by the release of plastic wastes into the environment, and proposed solutions for this problem. Chapter 3 takes a somewhat different approach, offering the views of 10 individuals as to their own particular experience with or thoughts about issues surrounding the use of plastics in everyday life.

The rest of the book is intended as a collection of resources for individuals who would like to learn more about the topic or would like to continue their own research on the topic. Chapter 4 provides brief sketches of individuals and organizations that have had or continue to have an important impact on the direction that plastics research has taken over the ages. Chapter 5 opens with a collection of basic data about the production, consumption, and disposal of plastic wastes in the United States and other parts of the world. It also contains selections from a variety of laws, administrative actions, recommendations, court cases, and other documents dealing with the role of plastics and plastic wastes in society.

Chapter 6 is a key feature of the book. It provides an annotated bibliography of books, articles, reports, and Internet sites dealing with a host of specific issues in the field of plastics pollution. The reader should keep in mind that the endnotes for chapters 1 and 2 also provide a rich collection of this type of resources. Chapter 7 offers a chronological history of major events in the history of plastics and microplastics. Finally, the glossary should be an invaluable aid to readers who would like a better understanding of some of the key terms that are used in discussions about plastics, microplastics, and related subjects.

# Plastics and Microplastics

## Introduction

How much plastic do you throw away, and how much do you recycle each day?

Here's an experiment you can do to find out:

- First, construct a table like the one shown in table 1.1. For the purpose of this experiment, the term "throw away" means depositing an object in any container that is *not* marked for recycling. For example, that might include a garbage can at home or a trash barrel at school that is *not* marked for recycling.
- Then make a list of all plastic products you have used over the past 24 hours. Some examples of common plastic products used in daily life are given in table 1.2.
- In column 2, estimate the weight of each item used. table 1.2 gives you a general idea as to how much some common plastic items weigh. Use a kitchen scale to weigh any other items on your list.
- In column 3, say what you did with the item when you were done with it. For example, you could say "trashed" or "recycled."
- Add up all the weights in column 2.
- Calculate the percentage plastic trashed and the percentage recycled.

---

The manufacture of a rubber tire, as of about 1928. (Library of Congress)

Table 1.1    A Summary of Your Plastic Use

| Item Used | Weight per Item (oz) | Trash/Recycle |
|---|---|---|
| | | |
| Total Weight of All Items | | 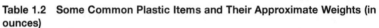 |
| Weight for Trash | | |
| Weight for Recycle | | |
| Percentage Trash | | |
| Percentage Recycle | | |

Table 1.2    Some Common Plastic Items and Their Approximate Weights (in ounces)

| Item | Approximate Weight* |
|---|---|
| coffee cup | 2 |
| water bottle | 1 |
| comb | 4 |
| toothbrush | 2 |
| detergent bottle | 2 |
| milk container | 3 |
| bubble wrap | varies |
| grocery bag (for shopping) | 4 |
| grocery bag (for products) | 0.2 |
| plastic sandwich wrap | 0.5 |
| trash bag | 3 |
| drinking straw | 0.1 |
| "peanut" packaging | varies |
| molded case (egg carton) | varies |
| electronics case | 4 |
| ball point pen | 3 |
| print cartridges | 20 |
| plastic tableware | 2 |
| plastic food container | 5 |
| disposable wipe | 2 |
| microwaveable food cover | 1 |

(continued)

Table 1.2    (*continued*)

| Item | Approximate Weight* |
|------|---------------------|
| plastic car/bike parts | varies |
| plumbing pipe | varies |
| squeeze bottle | 2 |
| disposable diaper | 3 |
| CD case | 3 |
| insulation materials | varies |
| Lid | 1 |
| pill bottle | 2 |
| garden hose | 16 |
| bucket and tub | 8 |
| plastic string and wire | 1 |
| plastic fibers and materials | varies |

*Weight varies depending on type of plastic, size of items, material used, and other factors.

Percentage = amount trashed or recycled / total amount of plastic used

If you don't want to go to the trouble of doing this experiment yourself, refer to table 1.3. That table is an example of the way Americans make use of plastics in their daily life.

What has this research taught you about the way you (and other young adults) use plastics in your daily life?

## Plastics in Daily Life

For virtually all of history, humans have relied on natural products and materials for the production of the "stuff" they needed and used in their daily lives. The earliest humans, for example, used rocks and stones to build their houses, native metals to make utensils and weapons, clay to make bowls and vases, animal horns as musical instruments, volcanic glass for cutting and jewelry, and oyster shells as utensils for food and decorative objects. Over time, humans became more creative in finding and modifying natural materials to meet their needs. For example, they invented boats made of reeds, writing objects made of goose quills, primitive forms of paper made from papyrus,

Table 1.3   A Typical Summary of a Young Adult's Plastic Use

| Item Used | Weight per Item (oz) | Trash/Recycle |
|---|---|---|
| water bottle | 1 (×3) = 3 | recycle |
| plastic sandwich wrap | 0.5 (×3) = 1.5 | trash |
| trash bag | 3 | trash |
| coffee cup | 2 | trash |
| ball point pen | 3 | trash |
| comb | 4 | trash |
| disposable wipe | 2 (×2) = 4 | trash |
| grocery shopping bag | 4 (×3) = 12 | trash |
| fresh vegetable bag | 0.2 (×4) = 0.8 | recycle |
| plastic fork | 2 | trash |
| plastic knife | 2 | trash |
| Total Weight of All Items | 37.3 oz. | |
| Weight for Recycle | 3.8 oz. | |
| Weight for Trash | 33.5 oz. | |
| Percentage Recycle | 10.2% | |
| Percentage Trash | 89.8% | |

bricks and mortar for construction of buildings, copper and bronze for the manufacture of weapons and tools, and flax for the making of cloth. The practice of using natural materials in a host of applications continues today. Most clothing comes from animals (wool) or plants (cotton); buildings are often constructed from marble, sandstone, or other kinds of rock; industrial equipment is usually manufactured from iron, steel, or other naturally occurring metals; a variety of glass materials come from sand, limestone, and other materials found in nature; leather is made from animal skins; and the world's activities are largely powered by fossil fuels, coal, oil, and natural gas.

All of these examples have one feature in common: they are all part of a natural cycle that exists, and has always existed, on earth. For example, objects made from papyrus or steel or wool or glass eventually wear out, catch fire, or break down by some other process. That process may take place over a few months or a few years. But it eventually occurs in some way or another. The chemical elements and compounds of which a material is made return to the land, water, or air and eventually reappear in some more complex form.

This process is illustrated by the carbon cycle, a series of events involving the element carbon. (For a diagram of the carbon cycle, see The Carbon Cycle 2007.) All living objects, as well as many rocks and minerals, contain carbon in some form or another. When that object dies or is destroyed, the carbon is released to the surrounding environment, often in the form of carbon dioxide gas. The carbon dioxide may then be dissolved in rivers, lakes, or ocean waters; absorbed by the earth; or, most commonly, released to the atmosphere. The carbon does not simply float away into space, lost to Earth forever. At the same time, new carbon does not just appear on Earth from some extraterrestrial source (with the rare exception of meteorites). The carbon that exists on Earth stays on Earth. It changes from one form to another, appearing in plants, animals, rocks, liquid compounds, and other forms. But it never just "goes away."

The carbon cycle is duplicated by similar processes for other elements and compounds. Water, for example, can occur in a liquid form, a solid form (ice and snow), and a gaseous form (water vapor). It constantly moves through the natural environment by means of evaporation, condensation, and other processes. The same occurs for all other elements and compounds. For some of the most important of these substances, such as nitrogen, phosphorus, and sulfur, natural cycles have been studied and described. But even the rarest of elements go through something of this kind. (For those interested in this topic, see Du and Graederl 2011.)

Plastics do not follow a cyclical pattern like that of all natural materials. Instead, they pass through a linear process that begins with their production from synthetic compounds (resins), such as ethylene, propylene, and styrene. Those resins are then used to make a variety of plastics such as polyethylene, polypropylene, and polystyrene. The plastics, in turn, become the raw material from which the countless commercial, industrial, domestic and other products are made. When those products wear out, or no longer serve a useful purpose, they meet one of three fates. They are deposited in some type of dump, such as a landfill; incinerated; or recycled. The recycling process might be considered a type of "plastic cycle" except for one important consideration: Recycling a plastic material more than once or twice is currently an economically unfeasible option. So it is generally more cost-effective just to throw a plastic away or burn it up than to break it down into its components and make new plastics out of those components.

## History of Plastics

Search the Internet for phrases such as "when did the age of plastics" begin, and you'll get a variety of dates and events to mark that period. One source, for example, says that the period began in 1939, with the display of plastic objects at the New York World's Fair, "The World of Tomorrow" (Taylor 2019). Other sources point to the invention of a tough, strong new material called Bakelite by Belgian inventor Leo Baekeland in 1907 (Murfin 2018). Still other observers suggest an even earlier date, 1869, when English chemist John Wesley Hyatt invented the first synthetic polymer, celluloid ("History and Future of Plastics" 2020).

The earliest period that can possibly be thought of as the "beginning" of plastics is the early-to-middle nineteenth century. Why then? Prior to that time, chemists devoted their studies entirely to nonliving materials, such as rocks, minerals, ores, gases, and the like. They spent their time taking these materials apart to determine their basic composition and finding ways to change them into new materials with different properties.

The reason for this line of research was a prevailing chemical (and philosophical) theory that said that living things were imbued with a special kind of "vital spirit" that directed their growth and development. There was no point in humans trying to study such materials, called organic because of their connection with plants and animals, as they had no control over these "vital spirits." (As poet Joyce Kilmer said many years later, "Only God can make a tree.")

That situation changed in 1828 when German chemist Friedrich Wöhler made a remarkable discovery. In his laboratory, he heated a sample of the mineral ammonium cyanate and found that it changed into an organic compound known as urea. No one questioned that ammonium cyanate was inorganic (it did not come from plants or animals) and that urea was organic (it is normally produced by the breakdown of proteins in the liver of animals). Although this experiment, all by itself, did not change the course of chemical science, it did mark the beginning of an entirely new view as to the differences between inorganic (nonliving) and organic (living) substances.

Wöhler's research became a spur for a host of similar experiments that marked the rise of modern organic chemistry. For our purposes here, the most important discovery of this period was that the defining characteristic of organic compounds was not their source in living organisms, but their chemical structure: They all contained the element carbon. Thus, over time, the term "organic chemistry" earned a new definition, one that holds today. Organic chemistry is the study of carbon compounds (a definition that has, however, a small number of important exceptions).

The key element in the rise of this new field of organic chemistry was the realization that materials previously thought to be "off-limits" to chemical research (e.g., the products of living organisms) could be studied, analyzed, and even replicated in much the same way as had rocks and minerals in earlier history. All of the components of the plastics industry eventually grew out of that realization.

## Natural and Synthetic Rubbers

For example, the one field of research that is sometimes cited as the beginning of polymer science is the study of natural rubber. Europeans had learned as early as the 1750s that indigenous peoples of Mesoamerica had developed methods for collecting sap (latex) from the Hevea brasiliensis (rubber) tree and converting it into hard balls used in their traditional games (Loadman n.d.-a). Very soon after this discovery was introduced to Europe, scientists and inventors began to look for practical applications of the products and ways of adapting its properties for additional uses. By 1770, for example, the noted English chemist Joseph Priestley is said to have discovered that the new material could be used to erase pencil marks on paper. He called the new product, therefore, rubber. Priestley apparently learned about the new material from the work of English inventor Edward Naime. In one of his books, Priestley notes that he had seen a small piece of material that could be used to erase marks made by a pencil on a piece of paper. The product was commercially available, he said, in cubical blocks about a half inch in size at a price of three shillings each (Priestley 1770; see footnote on page xv).

A few other inventions based on natural rubber were announced in the decades after Naime's work, most of which were of little or no value. One exception was the invention of a waterproof material invented by Scottish chemist Charles Macintosh in 1823. The material consisted of two pieces of cloth held together by a solution consisting of natural rubber dissolved in naphtha. Somewhat surprisingly, that invention remained popular through the ages and is still available in essentially its original form in various formats through today's United Kingdom-based Mackintosh company ("Charles Macintosh and His Famous Coat" n.d.).

One reason for the slow progress in the commercialization of natural rubber was its crucial physical properties. When made from latex at normal temperatures, it formed a hard, tough solid. As the temperature rose, however, it began to get soft and to melt. Imagine a set of automobile tires made out of that type of material! The solution to that problem, and the event marking the modern history of rubber products, occurred almost simultaneously in Great Britain and the United States in the late 1830s.

One fact that has intrigued historians of science for generations is the truism that some of the most important discoveries in science have been made in two or more places by different individuals at essentially the same time. Such was the case with the vulcanization of rubber. By the 1840s, the potential value of hardened rubber had become so apparent that several inventors were looking for a method to keep the product hard and solid when heated. One of the first of these individuals was American inventor Charles Goodyear. Goodyear went through a variety of chemicals that he thought might keep rubber in a solid form at high temperatures. He tried magnesium sulfate, calcium hydroxide, nitric acid, and other materials, without success. He finally found the right material, sulfur, but only in one of the most famous accidental discoveries in history. When sulfur was added to liquid rubber at room temperature, nothing happened to the rubber. But when Goodyear accidentally spilled some of the sulfur-rubber on top of his stove, he found that the product formed at exactly the properties for which he was looking: a resistance to melting when exposed to high temperatures. In early 1844, he applied for a patent for his new process, which he called vulcanization.

The only problem with this story was that a similar discovery had been made somewhat earlier in Great Britain by English inventor Thomas Hancock. Hancock had applied for a patent in Great Britain about eight weeks earlier than Goodyear did in the United States, and he eventually won that patent war. To add to the confusion (and ill feeling), a third person, English inventor Stephen Moulton, had also applied for and received a patent for the vulcanization process before Goodyear. In spite of this somewhat complex story, most historians today agree that Goodyear was the first person to discover vulcanization of rubber, and he is named the "Father" of that process. (See sketches of these three individuals and a summary of the controversy among them at Loadman n.d.-b.)

One important point about this whole discussion concerns what chemists actually "knew" about chemical compounds, their structures, and how they reacted with each other. Until the late 1910s, this "knowledge" was restricted to the macroscopic level, that is, the properties and changes that could be observed with one's eyes or a simple microscope. For example, researchers knew that natural rubber could be made into a solid, but that solid melted as it became warmer. Everyone could observe that information without any specialized equipment. But no one had the slightest idea what was going on with the atoms and molecules of which natural rubber is made. Today, one can say that those molecules at or below room temperature are long polymeric chains that remain in position relative to each other. At room temperature, that is, natural rubber is a stable solid. But raising the temperature of the material causes those molecules to begin moving about at higher rates of speed. Eventually, they move fast enough to lose any coherent structure and become, instead, a liquid.

Using this model, one can also explain how vulcanization solves the problem of natural rubber's softening at elevated temperature. The atoms that make up the chemical added to natural rubber, such as sulfur, form strong bonds between two adjacent rubber molecules, forcing them to remain in position even when the temperature rises. Today, any useful discussion of polymers and plastics takes into account the behavior of atoms and molecules of which these materials are made. That topic is in most cases, however, beyond the technical level presumed for this book. (A number of good resources on basic concepts in chemistry for young adults are available from a variety of sources. Web pages such as Bishop n.d. are one example.)

Attempts at making synthetic forms of natural rubber can be traced to the early 1920s, when several industries began to worry about the escalating price of the product. Since the vast majority of natural rubber came from and was controlled by nations in southeast Asia, there was, for many years, no options for countries elsewhere in the world. They paid the price asked for natural rubber or went without, an option that was not an option for developed nations worldwide.

Efforts to invent a synthetic form of natural rubber date to the late 1900s. In 1909, for example, a research team led by German chemist Fritz Hoffman at the Bayer Chemical company found a way to polymerize isoprene. Isoprene is the monomer found in natural rubber, one of whose technical name is, therefore, polyisoprene. Hoffman was able to design temperature, pressure, and catalytic conditions that made possible the conversion of isoprene to polyisoprene, similar in many ways to the natural product. Before long, researchers began to produce other types of synthetic rubber made from somewhat different monomers. The monomers were sometimes direct derivatives of isoprene, such as chloroprene, which has only one chlorine atom in place of one hydrogen atom. Another line of research explored the use of two different monomers for the production of a polyisoprene-like compound. The first successful attempt in this process was the invention of so-called Buna rubbers. That name comes from one of the monomers used in the process, butadiene, and the most common catalyst involved in the reaction, sodium (chemical symbol: Na).

In the late 1930s, the "option" of buying natural rubber or not from southeastern nations was no longer available. Those nations were among the first conquests made by the Japanese military at the outset of World War II. A critically important "hurry-up" campaign for the development of new methods for making synthetic rubber kicked into gear in the Allied nations. So successful was that campaign that by the end of the war the amount of synthetic rubber being produced worldwide was twice the amount available from natural sources ("United States Synthetic Rubber Program, 1939–1945" 1998). Since rubber is not technically a plastic, the topic of this book, readers interested in the history of natural and synthetic rubber are referred to other sources, such as Bisio 1990, De Guzman 2008, Herbert and Bisio 1985, Morton 1981, and "The Synthetic Rubber Project" 2015.

## Parkisine and Celluloid

The earliest plastic-like materials were based on naturally occurring polymers, almost always, cellulose. Of course, inventors knew nothing about the chemical structure, properties, and reactions associated with cellulose. But they were familiar with the many ways in which plant materials, such as hay, straw, grass, reeds, hemp, cotton, wood pulp, and tree limbs and branches, had been used over the centuries to make useful objects, construct buildings, and create a host of everyday products. So it is not surprising that researchers and inventors began to ask themselves how they could change the properties of plant material to make them stronger, longer lasting, more attractive, or better than the natural material in other ways. They tried a variety of procedures, such as trying out various solvents for the plant material, using different levels of heating, and adding other organic and inorganic materials to cellulose products.

Probably the first success in this field came in the mid-nineteenth century, when English inventor (and self-proclaimed chemist) Alexander Parkes invented a product he named after himself, Parkesine. It seems that Parkes had been motivated to follow this line of research by a letter from Swiss German chemist, Christian Schönbein, in 1846. Schönbein had apparently developed a method for converting paper into a solid, tough, transparent solid. Parkes tried a variety of chemicals as additives for his samples of cellulose, including sulfuric and nitric acid, vegetable oils, and other organic materials. The substance took on a doughy-like consistency that Parkes was able to mold and shape into various shapes. In 1862, he exhibited his products at the International Exhibition and was awarded a bronze medal for his work (Mossman 2017; a record of Parkes's original report on the material is available at Parkes 1866). Unfortunately, Parkes was not as good a businessman as he was an inventor, and the company he founded in 1866 to make and sell his product soon went out of business.

Parkes's research also led to the discovery of another early plastic, celluloid. Celluloid, in turn, had an earlier predecessor in a substance known as collodion. The discovery of collodion is murky, with credit for its invention being given to several researchers. (See Schönbein's own account of this controversy in Schoenbein 1849.) In one telling, the material was first discovered accidentally in 1846 by Swiss chemist Christian Friedrich Schönbein. He used his wife's cotton apron to mop up a spill consisting of sulfuric and nitric acids. He found that the cotton retained its appearance, but it caught fire and burned very quickly. The material soon become known as gun cotton and, as a derivative of nitric acid only, cellulose nitrate (also, nitrocellulose flash paper, flash cotton, flash string, and, in a somewhat modified form, pyroxylin). When dissolved in ether, Schönbein's discovery became a valuable medical tool. Applied to the skin, the ether evaporated rapidly, leaving behind a thin, tough covering that could remain place until healing had occurred.

In one respect, the true inventor of celluloid might be said to be Parkes, the creator of parkesine. Parkes had received more than 20 patents on products made from cellulose nitrate with a variety of solvents. One such product involved the use of an organic solvent called camphor. The mixture has a thick, syrupy appearance that dries to form a tough, strong solid. When Parkes was financially unable to proceed with the commercialization of his discoveries, English chemist John Wesley Hyatt assumed Parkes's patent for the process and is now generally regarded as the inventor of celluloid.

An important event in the popularization of celluloid occurred in 1863, when the billiard ball company of Phelan and Collender offered a $10,000 prize for the invention of a new type of billiard ball. The competition was created because the company was concerned that the traditional material from which billiard balls were made, ivory from the tusks of an elephant, was insufficient to meet the demands of the industry. Although the prize was never actually awarded, historians regard Hyatt's cellulose nitrate/camphor recipe actually met the conditions of the contest. He patented his own new discovery and went on to develop a highly profitable line of products from the substance that is now regarded as the first truly synthetic plastic (Boyd 2011; Robbett 2018).

## Bakelite

Bakelite was first patented by its creator, Belgian inventor, Leo Hendrik Baekeland, in 1907. At the time, Baekeland had already become a wealthy man because of his invention of another plastic material, Velox photographic paper, in 1893. As Baekeland continued his research into possible plastic materials, he decided to follow a line of research that had been suggested by German chemist Adolf von Baeyer and by his student, Werner Kleeberg, more than 30 years earlier. When von Baeyer and Kleeberg mixed two organic compounds, formaldehyde ($CH_2O$) and phenol ($C_6H_5OH$), with each other, a reaction occurred, producing a very hard, tough solid that could not be removed from their equipment. The two researchers could not imagine any commercial application for the procedure and did not proceed further with their studies of the reaction.

Baekeland repeated these experiments, modifying the conditions under which the reaction took place. He eventually found a way to stop the polymerization reaction before it was complete. At that point, the product was soluble in a variety of solvents, allowing changes to be made in the physical properties of the product with predetermined shapes for a host of purposes. In fact, Baekeland concluded that his discovery, which he named Bakelite, could be used to make "the thousand and one other articles of domestic and industrial utility now made of hard rubber, celluloid or kindred substances" (Baekeland 1910). In fact, Baekeland was not far off in his optimistic projections for the future of his invention. It was first adopted by the electrical and automotive industries because of its high resistance to electrical current, to heat, and to chemical action. It soon became popular for use in many domestic objects that had to be strong and resistant to damage in ordinary use, objects such as telephones; radios; cameras; electrical sockets; billiard balls, poker chips, chess sets, dominoes, and other table game equipment; kitchenware of many types; costume jewelry; musical instruments; gun parts; aerospace products; and all manners of toys.

(For images of some of the many objects made from Bakelite, see Heichelbech 2020; for a good review of Baekeland's work on the product, see Powers 1993.)

For the history of plastics, Baekeland's discovery was important because it was the first entirely synthetic plastic. Neither phenol nor formaldehyde is found in nature, so no natural products are used in the manufacture of the material. For a detailed discussion of the history, properties, and other relevant information about phenol formaldehyde plastics, see Fink 2018 (chapter 4).

## The Modern Age of Plastics

Baekeland's discovery of Bakelite marked the beginning of a modern age of plastics. It was followed over the next three decades by a flood of breakthroughs in the chemical nature of polymers and plastics, the discovery and invention of several important new monomers, and the creation of a host of polymers and plastics made from these monomers. Some of the highlights of this period are the following.

German chemist Fritz Klatte developed a method for reacting acetylene ($C_2H_2$) with acetic acid ($CH_3COOH$), resulting in the formation of the monomer vinyl acetate ($CH_3CO_2CH=CH_2$). He then was able to make polymers of the substance, called polyvinyl acetates, for which he received a patent in 1912. Polyvinyl acetates are still today a large and important class of plastics. Klatte is also credited with original research on the monomer vinyl chloride ($C_2H_3Cl$) and its polymerization to form polyvinyl chloride. The process he developed was not commercially feasible, however.

One of the world's most important and most popular plastic, polyethylene, was discovered by accident in 1933 by two researchers, Eric Fawcett and Reginald Gibson, at the Imperial Chemical Industries, in Northwich, England. While studying the reaction between ethylene and benzaldehyde in an autoclave, their device sprang a leak and the experiment had to be terminated. Upon investigation of the autoclave's interior, however, Fawcett and Gibson discovered a waxy, white material they found to be polyethylene. The company obtained a patent for the discovery three years later and began sales of a transparent film made of polyethylene in 1937. Today, that product is known as the "low-density" form of the polymer, LDPE (Jagger 2008).

In 1953, German chemist Karl Ziegler, and his graduate student, Erhard Holzkamp, found a way to produce a different, but equally important, form of polyethylene, high-density polyethylene (HDPE). For this discovery, Ziegler was awarded a share of the 1963 Nobel Prize in Chemistry. Today, more than 80 million tonnes (88 million short tons) of ethylene plastics of all kinds are produced today, making it the most popular of all kinds of plastics made worldwide.

Of at least equal importance as the discovery of new monomers and polymers during this period was research on the most fundamental facts as to how polymers and plastics form. Prior to the early twentieth century, researchers based their work almost entirely on intuition, good luck, imagination, and trial and error. The discovery of most early monomers and polymers, that is, was the consequence of someone's repeating an experiment over and over again, dozens or even hundreds of times to get positive results. A researcher would try various combinations of chemicals to start with, test their reaction under different conditions of temperature, pressure, types of catalysts, and other conditions. There existed no widely-accepted theoretical principles on which such research could be conducted.

That situation first began to change in the 1910s, largely as the result of the work of American chemist Gilbert Newton Lewis, with contributions by German chemist Walther Kossel and American chemist Irving Langmuir. In 1916, Lewis published a now-famous paper, "The Atom and the Molecule," in which he tried to explain how atoms combine with each other to form molecules. He suggested that two processes might occur. In one, one atom would give away one or more electrons to a second atom, forming a charged pair called an ionic bond. In the other process, two atoms would share one or more pairs of electrons with each other, a structure called a covalent bond. His ideas soon superseded the older view that chemical reactions were just a matter of different atoms mixing up with each other. It is, of course, the principle on which all modern bonding theory is based (Shaik 2006).

Lewis's connection to polymer science came through the work of German chemist Hermann Staudinger. Staudinger was very much interested in the chemical structure of very large molecules, molecules he called macromolecules. Although they were relatively well known to chemists, virtually nothing was known about their physical and chemical properties. In 1920, Staudinger published an important paper, "Über Polymerisation," in which he suggested that macromolecules were made of very simple compounds (monomers), joined to each other by covalent bonds. In his paper, Staudinger in fact introduced the term "polymerization" (Staudinger 1920). As with Lewis's research, Staudinger's studies were to provide a theoretical basis on which all future polymer research was to be based. For his discovery, he was awarded the Nobel Prize for Chemistry in 1953 ("The Foundation of Polymer Science by Hermann Staudinger (1881–1965)" 1999).

Over the last century, the discovery, invention, and production have continued apace. Some of the interesting discoveries include those of nylon (told earlier), Plexiglas, and Teflon. As with some other important breakthroughs in plastics research, the discovery of Plexiglas was largely the result of an accident. In the early 1930s, German chemist Otto Röhm was studying the properties of a new family of plastics, the poly(methyl methacrylates) (PMMAs). At one point, he placed a bottle containing the monomer on a table by a window. When sunlight began to pass through the window, the bottle broke apart, leaving behind a solid transparent material, PMMA. Later tests found the material to be at least as strong and tough as glass, resistant to acids and other chemicals, as well as to sunlight. By

1933, Röhm had obtained a patent for the new material, later marked under the brand names of Plexiglas, Lucite, Acrylite, and Perspex. ("This Is How Plexiglas® Came to Be" n.d.; slight variations on this story are available.)

And yet another important plastic discovered accidentally was Teflon, first produced in 1938 by American chemist Roy Plunkett and his colleagues at the Chemours Jackson Laboratory in New Jersey. Plunkett had been hired to study new types of refrigerants being developed that consisted of ethylene monomers in which one, two, three, or all four of the hydrogen atoms were replaced by either chlorine or fluorine atoms, or both. He began his studies with the monomer tetrafluoroethylene ($C_2F_4$). He stored several bottles of the monomer in a refrigerator, waiting to use them in his research. When he removed a bottle from the refrigerator and attempted to extract the gaseous tetrafluoroethylene, he found that nothing came out of the bottle. Yet the bottle weighed the same as when it was first placed in the refrigerator. He broke open the bottle and found a white, waxy solid coating the sides of the bottle. When he studied the properties of this material, he found one remarkable property: it would not stick to any material he tested. It also turned out to be chemically inert and heat resistant. Building on these properties, Plunkett developed a new type of plastic, polytetrafluoroethylene, now far better known as Teflon (Plunkett 2020).

Athletics has been one of the important influences on the invention and development of new plastics. All types of sporting equipment have long been made from one or another type of plastic. A somewhat different example is the use of plastics for playing fields, such as the construction of artificial ice rinks. As early as the 1960s, some inventors were using the newly discovered polyoxymethylene plastic discovered by the DuPont company a decade earlier as an artificial ice surface. These efforts went nowhere because a lubricant had to be added to the artificial ice continuously. Continued research eventually produced a more suitable product in the mid-1980s with the development of polyethylene surfaces (Chanda and Roy 2007, 7–46). The most popular product today is made of high-density polyethylene or ultrahigh molecular weight polyethylene that contains a lubricant, such as some type of silicone, embedded in the plastic. When a skater moves over the surface of this material, it releases small amounts of the lubricant, closely approaching the effects available with natural ice. Hundreds of artificial ice rinks are now in operation for personal recreation events or amateur or professional hockey leagues at locations around the world, many of which would otherwise not be able to support an ice rink (Eastaugh 2012; also compare this history with the history of artificial turf as in, e.g., Weeks 2015).

In some ways, the turning point in the modern history of plastics came with the onset of World War II. The demand for strong, sturdy, long-lived, dependable materials needed in every aspect of the military campaign, as well as the growing demands for domestic applications, created a huge increase in plastics production. As evidence of that fact, consider that the total worldwide output of all plastic materials in 1950 was only 2 million tonnes. That number grew slowly until the mid-1970s, before taking off to an estimated total of more than 380 million tonnes in 2015. Table 1.4 provides more data on this growth.

## Plastic Pollution

By the 1960s, the largely unmitigated enthusiasm for plastic products in the United States and around the world had begun to abate. Several reports of the entanglement of marine life in plastic products had begun to appear. And by the 1970s, reports of plastic pellets in ocean waters and on the ocean floor began to appear (Ryan 2015). Part of the concern over marine pollution reflected a more general developing concern among the general population of a host of environmental issues, often attributed to some degree to the publication of Rachel Carson's classic book, *Silent Spring* (Carson 1962). At least as important

**Table 1.4    Worldwide Production of Plastics, 1950–2015 (in millions of metric tonnes)**

| Year | Production |
|------|------------|
| 1950 | 2 |
| 1955 | 4 |
| 1960 | 8 |
| 1965 | 17 |
| 1970 | 35 |
| 1975 | 46 |
| 1980 | 70 |
| 1985 | 90 |
| 1990 | 120 |
| 1995 | 156 |
| 2000 | 213 |
| 2005 | 264 |
| 2010 | 313 |
| 2015 | 381 |

*Source*: Geyer, Roland, Jenna R. Jambeck, and Kara Lavender Law. 2017. "Production, Use, and Fate of All Plastics Ever Made." *Science Advances*. 3: e1700782. https://advances.sciencemag.org/content/advances/suppl/2017/07/17 /3.7.e1700782.DC1/1700782_SM.pdf, Table S1. Also note graphical representation of this trend in Figure S1.

was the continuing growth of plastic products and their popularity in industry, commerce, and daily life.

One outstanding example was the invention of the plastic bag in 1959 by Swedish engineer Sten Gustaf Thulin. Motivated by concerns over the loss of trees used in making paper bags, Thulin invented a similar bag made out of plastic, rather than paper. He envisioned his invention as a way that consumers could have a carry-all package that could be used over and over again. Of course, as it turned out, that was not the eventual fate of the plastic bag. Instead, it became an object used once and then thrown away. The easily recyclable paper bag was rapidly replaced by the nearly indestructible single-use plastic bag (Laskow 2014; Weston 2019).

It took more than half a century for the world's governmental agencies to appreciate the environmental hazards posed by these supposedly great improvements in domestic packaging.

Throughout the first two decades of the twenty-first century, data continued to roll in on the harm they (and other forms of plastic) were causing to the environment and human health. Finally, in the late 2000s, cities, states, and national government had begun to institute complete bans on the use of plastic bags or, less seriously, creating laws setting a cost for the use of such products. San Francisco was the first city in the United States to adopt such a ban in 2007. The first ban on plastic bags worldwide came in 2002 when Bangladesh adopted such a prohibition. That action came because of dramatic country-wide flooding of storm sewage systems caused by plastic bags during widespread flooding during the storm season ("Plastic Bag Bans and Bag Fees" 2020).

## Microplastics

A new field of plastics research has opened up in the past two decades: microplastics. The term "microplastics" does not refer to a specific kind of plastics, such as polyethylene or polystyrene. Instead, it refers to the size of a particle consisting of any type of plastic. Microplastics are most commonly defined as very small pieces of plastics less than 5 mm (millimeters; 0.2 inch) in diameter. Recently, an even smaller form of plastic has been defined, nanoplastics. There is no universally agreed-upon definition for the term, but it has generally been reserved for pieces of plastics that are between 1 and 1,000 nm (nanometers; billionths of a meter) in diameter (Gigault et al. 2018). Two other terms may also be used in plastic particles larger than microplastics: "mesoplastic" and "macroplastic." A mesoplastic particle is normally defined as one between 5 and 10 mm in diameter, while a macroplastic particle has a diameter of more than 20 mm in diameter. The term "macroplastic," therefore, refers to all of the visible forms of plastic that one encounters in everyday life, from discarded plastic bags to empty water bottles to pieces of plastic sheeting.

The study of microplastics dates only to about the early twenty-first century. When oceanographer and boat captain

Charles Moore first discovered the Great Pacific Garbage Patch in the Pacific Ocean in 1997, he noted that the vast majority of the garbage found there consisted of tiny pieces of plastic: microplastics. Moore's discovery inspired a burst of research on these previously unknown particles. The number of scientific papers on the effects of microplastics on aquatic organisms alone rose from zero in 2000 to more than 150 in 2017 (Paul-Pont et al. 2018). Researchers today continue to ask where microplastics are found, how did they get into the ocean or the air, what effects they have on marine organisms, how they might affect the health of humans and other animals, and related questions. Largely because the science is still young, much of even the most basic information about microplastics is still poorly understood (Waluda, Cavanagh, and Manno 2018).

## Primary Microplastics

Microplastics are generally classified in one of two categories: primary and secondary. This distinction is based on the mechanism by which the particles are formed. Primary microplastics are materials that have been made intentionally and specifically for some commercial purpose, of a specific size, normally less than 5 mm. The first patents for such materials were issued more than 50 years ago, and they have become only more popular over the decades. The amount of primary microplastics released to the environment today is estimated to be between 0.8 Mt and 2.5 Mt annually. A recent study has concluded that somewhere between 15 percent and 31 percent of all plastics found in the oceans today are primary microplastics (Boucher and Friot 2017).

The vast majority of primary microplastics are produced either for use in personal care and cosmetics products or for some industrial operation. The former products are generally classified as "wash off" or "leave on" products. Some examples of the former products are toothpaste, hair shampoo, shower gel, and shaving cream. The product may be used as an abrasive to improve cleaning of the skin or teeth, as an exfoliant to remove dead skin, or for some similar purpose. Examples of "leave on" products are deodorant, lipstick mascara, insect repellent, hair spray, and hair coloring. They are designed to modify or enhance one's person appearance or to provide some type of health screen. The plastic particles used for these purposes are generally known as *microbeads*.

Primary microplastics also have a number of industrial uses, one of the most important of which is as an abrasive. The tiny plastic particles added to a spray are able to remove surface irregularities and achieve a smooth covering. They can also be added to paints and coatings to improve a material's adherence properties ("Intentionally Added Microplastics in Products" 2017). The amount of microplastic in any particular product can range widely, anywhere from 1 to 90 percent of the product's weight. The amount of microplastic in a shower gel, for example, can be equal to that of the packaging in which it came ("Plastic in Cosmetics" n.d.) Some of the specific uses for microplastics, as well as the plastics from which they come, is shown in table 1.5.

As important as these sources of microplastics may be, the large contributor to the release of those materials to the environment are two routine human activities: washing clothes and driving a vehicle. Studies show that two-thirds of all microplastic particles recovered from the environment came originally from the laundering of synthetic textiles and the erosion of plastic materials from vehicle tires. An estimated 34.8 percent of all microplastics came from the first of these sources and about 28.3 percent from the second source (Boucher and Friot 2017, 21). Data on this issue are still somewhat sparse, but one study has found that the laundering of a single load of synthetic textiles weighing about 12 pounds results in the loss of about 700,000 individual microplastic particles (Napper and Thompson 2016; note that products from textiles and tires are sometimes classified as secondary microplastics; see, e.g., "Denmark Identifies Main Sources of Microplastic Pollution" 2015).

Table 1.5   Some Applications of Microplastics

| Use | Plastic(s) |
| --- | --- |
| Bulking | Nylon-6 |
| | Nylon-12 |
| | Poly(ethylene isoterephthalate) |
| | Polypropylene |
| | Polytetrafluoroethylene (Teflon) |
| Viscosity control | Nylon-6 |
| | Nylon-12 |
| | Poly(butylene terephthalate) |
| | Polyethylene |
| | Polypropylene |
| | Polyacrylate |
| | Ethylene/propylene/styrene copolymer |
| | Butylene/ethylene/styrene copolymer |
| Abrasive | Polyethylene |
| Film formation | Poly(butylene terephthalate) |
| | Poly(ethylene terephthalate |
| | Poly(pentaerythrityl terephthalate) |
| | Polystyrene |
| | Polyurethane |
| | Ethylene/methylacrylate copolymer |
| | Trimethylsiloxysilicate (silicone resin) |
| Hair fixative | Poly(ethylene terephthalate) |
| | Acrylate copolymers |
| | Allyl stearate/vinyl acetate copolymers |
| Aesthetic agent | Poly(ethylene terephthalate) |
| | Styrene acrylates copolymer |

Source: "Plastics in Cosmetics." n.d. United Nations Environment Programme, 2. http://wedocs.unep.org/bitstream/handle/20.500.11822/21754 /PlasticinCosmetics2015Factsheet.pdf.

The loss of microplastics to the environment through the types of products described earlier can be classified as "unintentional"; that is, manufacturers did not design a product with the goal of it being released to the environment (the exception being certain cosmetic products, such as toothpaste, when that is, indeed, the purpose of the product). About 85 percent of all the microplastics released to the environment come from personal care and cosmetic products. Other products and practices are also responsible for the unintentional release of microplastics to the environment. The most common of those practices is laundering of synthetic textiles, a process during which tiny pieces of plastic break off a textile and are carried away with waste water. This process is thought to account for about 12 percent of all microplastics in the environment. Yet another process is vehicular traffic, in which small bits of tires break off and are washed away into rivers, lakes, and, ultimately, the oceans (Boucher and Friot 2017).

### Secondary Microplastics

Anyone who has crushed a piece of plastic made from Styrofoam knows how easily it breaks down into small pieces of plastic. If that process were to be continued over and over again, the Styrofoam would break down into smaller and smaller pieces until they reach the size of microplastic. Microplastic that is formed from the breakdown of larger pieces of plastic is known as secondary microplastic. Secondary microplastics are the materials we are most likely to hear about in discussions of microplastic pollution. They are formed primarily by photodegradation (exposure to sunlight) and erosive actions such as wind or wave motion on familiar plastic products, such as bottles and other containers, straws and eating utensils, fishing nets, plastic cups and lids, sheets of plastics, and all kinds of plastic packaging materials.

Most studies show that secondary microplastics occur at a substantially higher rate in the natural environment than do primary microplastics. A landmark study by the United Nations Environment Programme placed the relative contributions of secondary and primary microplastics worldwide at about 69–85 percent for the former and 15–31 percent for the latter (Boucher and Friot 2017, 5). This comparison differs substantially from region to region across the planet. In Africa and the Middle East, for example, the relative quantity of secondary and primary microplastics was 1.53 Mt per year to 0.13 Mt per year. In China, that ratio was 2.21 Mt per year to 0.24 Mt per year. Only North America shows a reverse pattern, with the ratio of secondary to primary microplastics at

0.07 Mt per year to 0.26 Mt per year (Boucher and Friot 2017, Figure 10, 28).

As noted earlier, some authorities classify microparticles produced by "wear and tear," such as the laundering of clothes and friction between the road and the tires on a car, as a form of secondary, rather than primary, production. No widespread agreement has been reached on this point.

## Transport of Microplastics

Many experts have begun to speak and write about microplastics in terms of a "plastic cycle" (Horton and Dixon 2018). That term is somewhat similar to the carbon cycle, water cycle, nitrogen cycle, and other cycles discussed earlier. One major difference with regard to the "plastic cycle" is that it does not occur in nature. It exists only because of the activities of human production processes. Another difference is that the "plastic cycle" is not really a cycle; it does not consist of a continuous flow of plastic materials from production to transport to use to disposal and back to production. It is a linear process that ends with disposal (except in those instances in which plastic materials are successfully recycled several times).

### Land and Air Deposition

Microplastics are produced at every stage of that cycle. In some cases, production is intentional, as in the production of the very small particles (*nurdles*) used as raw materials in the manufacture of plastics. In other cases, microplastics are lost along the way as an unintentional consequence of some step in the plastic cycle. The first instance in which that happens is during the production of plastics and microplastics in manufacturing operations. The nurdles from which plastics are made are subject to all kinds of damage during the production process. They can, when sold as nurdles, have several types of flaws, such as broken pieces, undersized pellets, or smashed pieces. They can also exist in unacceptable forms, such as dust, ribbons, or crumbs, none of which can be used in the production process. These damaged pieces then become part of the unintentional loss of microplastics to the environment. They may just escape during the production stage, or they may be captured and sent to some type of disposal system. Flawed nurdles can also be produced by operations associated with the processing, storage, transport, and other treatment of microplastics, such as loading and unloading the material, storage in a silo or other large container, packaging for shipment, and blending and feeding the nurdles (Dhodapkar, Trottier, and Smith 2009). The total amount of microplastics released to the environment as a result of production errors such as these has been estimated at 400 tonnes in the European Union alone in one year (Sundt, Schulze, and Syversen 2014).

What happens to microplastics once they have been produced during production or use, as described earlier? Perhaps the most frequently mentioned destination for these microplastics is the oceans, where they have become widely recognized as one of the most serious marine pollutants worldwide today. But microplastics have been found to be a significant pollution in soils and the air also.

Waste treatment systems have been devised for the treatment or removal of pollutants in both the air and in water. These systems are largely successful in preventing toxic solids, liquids, and gases from escaping into the environment. They are generally not very successful in capturing microplastics, however, largely because of the small size of microbeads, nurdles, and other microplastic particles. Those materials largely slip through filters and other devices that are standard parts of air and water waste treatment facilities.

One of the most interesting developments in this field is that microplastics may be at least as important, and possibly much more important, as a pollutant on land as in the oceans. A major reason for this fact is that microplastics released during industrial processes and in municipal solid waste systems end up in the solid residue collected in these systems, the so-called *sludge* found as an end product of waste treatment plants. One of the most common uses for this sludge is as a fertilizer for commercial agricultural operations. Microplastics in the sludge fertilizer are absorbed by the soil and may become part of the plants grown on those lands.

Several studies have now shown that farmlands fertilized by sludge in the European Union now contain measurable and possibly significant levels of microplastics. One such study estimated that somewhere between 125 and 850 tons of microplastics per million population are added annually in various parts of Europe (Nizzetto, Futter, and Langaas 2016). Other research shows that microplastics may remain in the soil for at least 15 years after their original application and can migrate to a depth of at least 25 centimeters (ten inches) (Zubris and Richards 2005; also see Corradini, et al. 2019; for a good review of this topic, see Horton and Dixon 2018).

Atmospheric fallout may also be a significant source of microplastics on land. The logic behind this conclusion is that several events occurring in urban areas primarily, but also in some suburban areas, may result in the release of microplastics to the air. The release of microparticles from tires is one obvious example of this event. Since microplastics are so small and so light they would presumably be distributed over wide regions by wind and precipitation. So far, however, there has been relatively little research to confirm this presumption. One study that claims to be the first of its kind in this area was reported in 2016. That year-long study was conducted in several parts of Paris, and it found that anywhere from 2 to 355 microparticles of plastic per square meter per day were observed. About 29 percent of these microparticles came from textiles and other synthetic materials. These results would suggest an annual release of 3–10 tons of microplastics per year over a land area of 2500 square kilometers (about 960 square miles) (Dris et al. 2016).

*Water Deposition*

The path taken by microplastics to the oceans is a relatively simple and straightforward one. No matter what their source, microplastics eventually make their way into lakes, rivers, streams, and other bodies of water. Even if they are deposited in the soil for some period of time, they eventually are carried away by the movement of groundwater into nearby lakes and rivers. If they are released to the atmosphere, they are eventually returned to earth's surface by rain, snow, sleet, or some other form of precipitation. They are then carried into rivers and lakes before being transported to the oceans. Recent research suggests that more than 90 percent of all plastic wastes are transmitted by only 10 large rivers around the world. Primary among those rivers are the Yangtze River (Chang Jiang, 1,469,481 tons per year), Indus River (164,332 tons per year), Yellow River (Huang He, 124,249 tons per year), and Hai He (91,858 tons per year) ("Our Planet Is Drowning in Plastic Pollution" n.d.). So it's little wonder that so much attention has been paid by researchers to the role played by microplastics in the oceans. Maritime activities are also responsible for the deposition of microplastics in the oceans. These activities include wear and tear from fishing nets, ropes, pots, and other fishing equipment as well as intentional disposal of or accidental loss of fishing equipment at sea (Lusher, Hollman, and Mendoza Hill 2017).

Reliable data on the amount of microplastics in the oceans is sparse. Until recently, researchers placed the amount of microplastics in the seas at about 10 microfragments per cubic meter of ocean water. More recent research has, however, raised that number substantially, to as much as 8.3 million pieces of microplastic per cubic meter. This change has been due to a large extent because traditional methods of capturing microplastic particles did not produce accurate results. The size of holes in nets used to capture microplastics was large enough that large numbers of those particles were not captured and, therefore, not measured (Brandon, Freibott, and Sala 2019).

Marine microplastics tend to occur in one of two locations: along the ocean surface or buried in sediments on the ocean bottom. This separation of microparticles is a consequence of the relative density of various types of plastics. For example, the densities of polyethylene and polypropylene are just less than 1.00 gm/cm$^3$ (one gram per cubic centimeter; 0.91-0.95 gm/cm$^3$ and 0.90-0.92 gm/cm$^3$, respectively). These particles will, therefore, float on seawater (whose density is 1.027 gm/cm$^3$) and contribute to its surface sediment. The densities of most other plastics are greater than 1.00 gm/cm$^3$, such that these microparticles will sink to the bottom of the ocean. For example, densities of polystyrene, polyvinyl chloride, and polyethylene terephthalate are, respectively, 1.04–1.09 gm/cm$^3$, 1.16–1.30 gm/cm$^3$, and 1.34-1.39 gm/cm$^3$ (Lusher, Hollman, and Mendoza Hill 2017, Table 3.2, 26).

Experts have suggested varying estimates for the amount of microplastic in the world's oceans. In one study, for example, the estimate was five trillion pieces of plastics with a total mass of 250,000 tons "afloat on the sea." No estimate of plastic wastes in bottom sediments was provided (Eriksen et al. 2014). Another study placed those numbers at between 15 and 51 trillion pieces of microplastic, with a total mass of between 93 and 236 thousand metric tons (Van Sebille et al. 2015). (These estimates are based on different methodologies, different locations, different definitions of "plastic wastes," and other variables.) Several studies have taken note of an irregular distribution of microplastics at various depths in the oceans. Many particles are found or near the water's surface (2–4 particles per cubic meter), with a maximum concentration at about 200 meters below sea level (10–12 particles per cubic meter). The concentration of particles then falls off to about 3 particles per cubic meter below 1,000 meters (Choy et al. 2019).

It seems likely that virtually all microplastics will eventually end up in bottom sediments, the final sink for the materials. Even microplastics originally found in surface sediments are likely to become more dense by the accumulation of external biolayers, causing them to sink to the ocean bottom. Numerous studies have been conducted on this pattern with the general result that the number of microparticles in the bottom sediment tends to decrease with the depth of the sediment. The explanation for this phenomenon is probably that the production of plastics over time has increased substantially, resulting in an increase in plastic wastes (and, therefore, microplastics) in the ocean (Maes et al. 2017).

Having set the scene for the rise of plastics in the United States and throughout the world since the mid-nineteenth century, we now turn in chapter 2 to some of the many serious problems these new wonder products have led to, and continue to lead to, in many aspects of human life. In that chapter, we also review some of the solutions that have been suggested for this problem.

## References

Baekeland, Leo H. 1910. "Bakelite, A New Insulating Material." *Engineering Magazine*. 38: 910–913. https://books.google.com/books?id=7BHOAAAAMAAJ&pg=PA910.

Bishop, Cherelle. n.d. "What Is a Polymer?" https://www.teachengineering.org/content/csu_/lessons/csu_polymer/csu_polymer_lesson01_polymer_presentation_tedl_mhf.pdf.

Bisio, Attillo. 1990. *Synthetic Rubber: The Story of an Industry*. Houston, TX: International Institute of Synthetic Rubber Producers.

Boucher, Julien, and Damien Friot. 2017. *Primary Microplastics in the Oceans: A Global Evaluation of Sources*. Gland, Switzerland: IUCN. https://portals.iucn.org/library/sites/library/files/documents/2017-002-En.pdf.

Boyd, Jane E. 2011. "Celluloid: The Eternal Substitute." Distillations. https://www.sciencehistory.org/distillations/celluloid-the-eternal-substitute.

Brandon, Jennifer A., Alexandra Freibott, and Linsey M. Sala. 2019. "Patterns of Suspended and Salp Ingested Microplastic Debris in the North Pacific Investigated with

Epifluorescence Microscopy." *Limonology and Oceanography Letters.* 5(1): 46–53. https://doi.org/10.1002/lol2.10127.

"The Carbon Cycle." 2007. UCAR Center for Science Education. https://scied.ucar.edu/carbon-cycle.

Carson, Rachel. 1962. *Silent Spring.* Boston: Mifflin.

Chanda, Manas, and Salil K. Roy. 2007. *Plastics Technology Handbook.* 4th ed. London: Taylor & Francis.

"Charles Macintosh and His Famous Coat." n.d. Google Arts & Culture. https://artsandculture.google.com /theme/charles-macintosh-and-his-famous-coat /SwLSEIbM4qLrIg.

Choy, C. Anela, et al. 2019. "The Vertical Distribution and Biological Transport of Marine Microplastics Across the Epipelagic and Mesopelagic Water Column." *Scientific Reports.* 9(1). https://www.nature.com/articles/s41598-019 -44117-2.

Corradini, Fabio, et al. 2019. "Evidence of Microplastic Accumulation in Agricultural Soils from Sewage Sludge Disposal." *Science of the Total Environment.* 671: 411–420. https://reader.elsevier.com/reader/sd/pii /S004896971931366X.

De Guzman, Dora. 2008. "History of the Synthetic Rubber Industry." Independent Commodity Intelligence Services. https://www.icis.com/explore/resources/news/2008/05/12 /9122056/history-of-the-synthetic-rubber-industry/.

"Denmark Identifies Main Sources of Microplastic Pollution." 2015. Chemical Watch. https://chemicalwatch.com /43667/denmark-identifies-main-sources-of-microplastic -pollution.

Dhodapkar, Shrikant, Remi Trottier, and Billy Smith. 2009. "Measuring Dust and Fines in Polymer Pellets: The Ability to Carry out Such Measurements Can Help Operators Improve Quality Control, Assess Equipment Performance and Optimize the Process." *Chemical Engineering.* 116(9): 24–29.

Dris, Rachid, et al. 2016. "Synthetic Fibers in Atmospheric Fallout: A Source of Microplastics in the Environment?" *Marine Pollution Bulletin.* 104(1–2): 290–293. https://doi .org/10.1016/j.marpolbul.2016.01.006.

Du, Xiaoyue, and T. E. Graederl. 2011. "Uncovering the Global Life Cycles of the Rare Earth Elements." *Scientific Reports.* 1: 145. https://doi.org/10.1038/srep00145.

Eastaugh, Natalie. 2012. "Synthetic Ice Rinks for All Times of the Year." Ezine Articles. https://ezinearticles.com /?Synthetic-Ice-Rinks-For-All-Times-of-the-Year&id =6916012.

Eriksen, Marcus, et al. 2014. "Plastic Pollution in the World's Oceans: More than 5 Trillion Plastic Pieces Weighing over 250,000 Tons Afloat at Sea." *PLOS One.* https://doi.org/10 .1371/journal.pone.0111913.

Fink, Johannes Karl. 2018. "Phenol/Formaldehyde Resins." In Johannes Karl Fink, ed. *Reactive Polymers Fundamentals and Applications: A Concise Guide to Industrial Polymers.* Amsterdam, The Netherlands: William Andrew.

"The Foundation of Polymer Science by Hermann Staudinger (1881–1965)." 1999. American Chemical Society and Gesellschaft Deutscher Chemiker. https://www.acs.org /content/acs/en/education/whatischemistry/landmarks /staudingerpolymerscience.html.

Gigault, Julien, et al. 2018. "Current Opinion: What Is a Nanoplastic?" *Environmental Pollution.* 235: 1030–1034.

Heichelbech, Rose. 2020. "13 Things You Won't Believe Were Once Made from Bakelite." Dusty Old Thing. https:// dustyoldthing.com/13-things-bakelite/.

Herbert, Vernon, and Attilio Bisio. 1985. *Synthetic Rubber: A Project That Had to Succeed.* Westport, CT: Greenwood Press.

"History and Future of Plastics." 2020. Science History Institute. https://www.sciencehistory.org/the-history-and -future-of-plastics.

Horton, Alice A., and Simon J. Dixon. 2018. "Microplastics: An Introduction to Environmental Transport Processes." *Wiley Interdisciplinary Reviews: Water.* 5(2): https://doi.org /10.1002/wat2.1268.

"Intentionally Added Microplastics in Products." 2017. Amec Foster Wheeler Environment & Infrastructure UK Limited. https://ec.europa.eu/environment/chemicals/reach /pdf/39168%20Intentionally%20added%20microplastics %20-%20Final%20report%2020171020.pdf.

Jagger, Anna. 2008. "Polyethylene: Discovered by Accident 75 Years Ago." Independent Commodity Intelligence Services. https://www.icis.com/explore/resources/news /2008/05/12/9122447/polyethylene-discovered-by -accident-75-years-ago/.

Laskow, Sarah. 2014. "How the Plastic Bag Became so Popular." The Atlantic. https://www.theatlantic.com /technology/archive/2014/10/how-the-plastic-bag-became -so-popular/381065/.

Loadman, John. n.d.-a "Charles Marie de la Condamine." Everything You Wanted to Know about Rubber. http:// www.bouncing-balls.com/.

Loadman, John. n.d.-b "Natural Rubbers—The People." Everything You Wanted to Know about Rubber. http:// www.bouncing-balls.com/.

Lusher, Amy, Peter Hollman, and Jeremy Mendoza-Hill, 2017. "Microplastics in Fisheries and Aquaculture: Status of Knowledge on Their Occurrence and Implications for Aquatic Organisms and Food Safety." Food and Agriculture Organization of the United Nations. http://www.fao.org/3 /a-i7677e.pdf.

Maes, Thomas, et al. 2017. "Microplastics Baseline Surveys at the Water Surface and in Sediments of the North-East Atlantic." *Frontiers in Marine Science*. 4: 1–13. https://doi.org/10.3389/fmars.2017.00135.

Morton, Maurice. 1981. "History of Synthetic Rubber." *Journal of Macromolecular Science: Part A—Chemistry*. 15(7): 1289–1302.

Mossman, Susan. 2017. "Early Plastics: Perspectives 1850–1950." *Ferrum*. 89: 14–24. This article is available online, but only by searching for the author's name and article title. The author has also edited a book on the same topic, *Early Plastics: Perspectives 1850–1950*. 2000. London: Continuum International Publishing Group.

Murfin, Patrick. 2018. "Bakelite Ushers in a New Era—The Age of Plastic." Hectic, Rebel, a Thing to Flout. https://patrickmurfin.blogspot.com/2018/02/bakelite-ushers-in-new-era-age-of.html.

Napper, Imogen E., and Richard C. Thompson. 2016. "Release of Synthetic Microplastic Plastic Fibres from Domestic Washing Machines: Effects of Fabric Type and Washing Conditions." *Marine Pollution Bulletin*. 112(1–2). https://doi.org/10.1016/j.marpolbul.2016.09.025.

Nizzetto, Luca, Martyn Futter, and Sindre Langaas. 2016. "Are Agricultural Soils Dumps for Microplastics of Urban Origin?" *Environmental Science & Technology*. 50(20): 10777–10779. https://doi.org/10.1021/acs.est.6b04140.

"Our Planet Is Drowning in Plastic Pollution." n.d. United Nations Environment Programme. https://www.unenvironment.org/interactive/beat-plastic-pollution/.

Parkes, Alex. 1866. "On the Properties of Parkesine and its Application to the Arts and Manufactures." *Journal of the Franklin Institute*. 81(6): 384–388. https://books.google.com/books?id=Lt85AQAAMAAJ&pg=PA264.

Paul-Pont, Ika, et al. 2018. "Constraints and Priorities for Conducting Experimental Exposures of Marine Organisms to Microplastics." *Frontiers in Marine Science*. 5. https://doi .org/10.3389/fmars.2018.00252.

"Plastic Bag Bans and Bag Fees." 2020. Surfrider Foundation. https://www.surfrider.org/pages/plastic-bag-bans-fees.

"Plastic in Cosmetics." n.d. United Nations Environment Programme. http://wedocs.unep.org/bitstream/handle/20 .500.11822/21754/PlasticinCosmetics2015Factsheet.pdf.

Powers, Vivian. 1993. "Leo Hendrick Baekeland and the Invention of Bakelite." American Chemical Society. https:// www.acs.org/content/acs/en/education/whatischemistry /landmarks/bakelite.html#invention-of-bakelite.

Priestley, Joseph. 1770. *Familiar Introduction to the Theory and Practice of Perspective*. London: J. Johnson & J. Paine. https://archive.org/details/afamiliarintrod00conggoog.

Robbett, Mary Kate. 2018. "Imitation Ivory and the Power of Play." Lemelson Center for the Study of Invention and Innovation. https://invention.si.edu/imitation-ivory-and -power-play.

"Roy J. Plunkett." 2020. Science History Institute. https:// www.sciencehistory.org/historical-profile/roy-j-plunkett.

Ryan, Peter J. 2015. "A Brief History of Marine Litter Research." In Melanie Bergmann, Lars Gutow, and Michael Klages, eds. *Marine Anthropogenic Litter*. Cham, Switzerland: Springer International Publishing, 1–25.

Schoenbein, C. F. 1849. "On Ether Glue or Liquor Constringens; and Its Uses in Surgery." *The Lancet*. 1: 289–290. https://books.google.com/books?id =ORZAAAAcAAJ&pg=PA289.

Shaik, Sason. 2006. "The Lewis Legacy: The Chemical Bond—A Territory and Heartland of Chemistry." *Journal of Computational Chemistry*. 28(1): 51–61. https:// onlinelibrary.wiley.com/doi/pdf/10.1002/jcc.20517.

Staudinger, H. 1920. "Über Polymerisation" (in German). *Berichte Der Deutschen Chemischen Gesellschaft (A and B Series)*. 53(6): 1073–1085. https://onlinelibrary.wiley.com /doi/epdf/10.1002/cber.19200530627.

Sundt, Peter, Per-Erik Schulze, and Frode Syversen. 2014. "Sources of Microplastic Pollution to the Marine Environment." Mepex. https://www.miljodirektoratet.no /globalassets/publikasjoner/M321/M321.pdf.

"The Synthetic Rubber Project." 2015. Science Reference Services. Library of Congress. http://www.loc.gov/rr/scitech /trs/trschemical_rubber.html.

Taylor, David A. 2019. "This Is How the Age of Plastic Began." Mother Jones. https://www.motherjones.com /environment/2019/03/this-is-how-the-age-of-plastics -began/.

"This Is How Plexiglas® Came to Be." n.d. Röhm. https:// www.world-of-plexiglas.com/en/the-history-of-plexiglas -this-is-how-it-came-about/.

"United States Synthetic Rubber Program, 1939–1945." 1998. American Chemical Society. https://www.acs.org /content/acs/en/education/whatischemistry/landmarks /syntheticrubber.html.

Van Sebille, Erick, et al. 2015. "A Global Inventory of Small Floating Plastic Debris." *Environmental Research Letters*. 10(12): 124006. https://iopscience.iop.org/article/10.1088 /1748-9326/10/12/124006/pdf.

Waluda, C. M., R. D. Cavanagh, and C. Manno. 2018. "A Cross-Sectoral Approach to Tackle Ocean Plastic Pollution." *Eos*. 99. https://doi.org/10.1029 /2018EO107159.

Weeks, Jennifer. 2015. "Turf Wars." Science History Institute. https://www.sciencehistory.org/distillations/turf-wars.

Weston, Phoebe. 2019. "Plastic Bags Were Created to Save the Planet, Inventor's Son Says." *Independent*. https://www

.independent.co.uk/environment/plastic-bags-pollution
-paper-cotton-tote-bags-environment-a9159731.html.

Zubris, Kimberly Ann V., and Brian K. Richards. 2005.
"Synthetic Fibers as an Indicator of Land Application of
Sludge." *Environmental Pollution.* 138(2): 201–211. http://
cwmi.css.cornell.edu/Sludge/Synthetic.pdf.

# 2 Problems, Controversies, and Solutions

## Introduction

Human life in the twenty-first century is subsumed in a world of plastics. From the tiniest bits of microplastic used to improve the efficiency of toothpaste to the largest pieces of plastic employed for the construction of miles-long gas pipelines, plastics appear everywhere in a person's life. An interesting fact about this situation is that the growth of the plastics industry did not really occur as an effort to meet consumer demands in a wide range of fields. Indeed, developments in plastics began just at a period in the life of Americans (the end of the Great Depression) when the philosophy of save and reuse had reached its peak. Instead, the surge of plastic products came largely because of the enthusiasm of professional chemists and industrialists for this seemingly miracle product. Consumers were encouraged to scrap the notion of reusing materials and products in their everyday life and, instead, to use it once and throw it away. Indeed, one of the classic examples of that drive was a *Life* magazine spread in 1955 showing a family tossing plastic napkins, plates, utensils, and other products as a lead-in to its article on "Throwaway Living" (Gilmore 2019).

As noted in chapter 1, however, that lifestyle appears to have no end. New kinds of plastics with a host of new uses are being produced every year. The "throwaway" mentality has led to a massive solid waste problem that grows substantially every year.

---

A torn plastic bag drifts over a tropical coral reef causing a hazard to marine life. (Whitcomberd/Dreamstime.com)

Instead of extolling the joys of ever more new plastic invention, many experts and everyday citizens are now beginning to ask how humans can deal with the host of environmental and health problems created by the surplus of plastic wastes and how the continued growth of this problem can be contained. This chapter reviews some of the most pressing issues relating to the manufacture, use, and disposal of plastic in today's world.

## Plastics Production

The growth in the production of plastics over the past 70 years has been astounding. According to what may be the best data on the topic currently available, the annual global production of all resins, fibers, and additives increased from about 2 Mt (million metric tonnes) in 1950 to 380 Mt in 2015 (Geyer, Jambeck, and Law 2017; unless otherwise noted, all data in this section come from this source). The total quantity of plastics produced during this period is estimated to be about 7,800 Mt, half of which has come in the last 13 years. By far, the largest use for these plastics has been in the field of packaging (an estimated 162.6 Mt in 2018), followed by building and construction (72.8 Mt), textiles (67.0 Mt), consumer and institutional products (45.9 Mt), and transportation (30.0 Mt). The plastic made in the largest quantity worldwide in 2015 was all forms of polyethylene (36% of all nonfiber plastics), followed by polypropylene (21%) and polyvinyl chloride (12%). (Data for fiber plastics are somewhat more difficult to obtain and are less reliable. Geyer, Jambeck, and Law estimate that the polyesters, polyamide, and acrylic plastics used in making fibers constitute about 12 percent of all plastic production today.)

Among all raw materials used in the plastics industry, ethylene is produced in the largest volume of any other compound. In fact, it is the organic compound produced in the largest volume of any chemical in the world (Milner 2017). Exact numbers are difficult to obtain, but estimates suggest that about

134 Mt of ethylene was produced worldwide in 2014, with about 25 Mt of the compound being produced in the United States for that year ("Ethene (Ethylene)" 2020, "Ethylene Production in the United States from 1990 to 2018" 2020). Propylene is also among the top 10 chemicals produced both worldwide and in the United States. Every indication suggests that ethylene and propylene production are likely to continue to grow in the foreseeable future (Boswell 2019).

## Future Trends in Plastics Production

One of the most famous quotations in the history of motion pictures comes from the 1967 United Artist film, *The Graduate*, starring Dustin Hoffman as Benjamin Braddock. Braddock is a recent college graduate being counseled by an older friend of the family and successful businessman, Mr. McGuire. In the historic exchange, McGuire says to Benjamin,

> "I want to say one word to you. Just one word."
> Benjamin replies: "Yes, sir."
> Mr. McGuire continues: "Are you listening?"
> And Benjamin answers: "Yes, I am."
> Mr. McGuire then provides his advice for a successful business career in one word: "Plastics." ("The Graduate Script—Dialogue Transcript" n.d.)

Mr. McGuire's advice was based on the fact that the plastics industry was a new and growing business opportunity that held great promise for any young college graduate. The outlook for the plastics industry in 2021 is at least as bright, if not more so, than it was in 1967. Today, that promise arises because of growing interest in traditional plastics, as well as a host of new technologies and applications for the materials. For example, great strides are being made in the development of so-called smart plastics. These materials are able to change their size, shape, and physical properties when exposed to appropriate stimuli. Of special interest to individuals both within and outside the plastics industry is the topic of biodegradable plastics (or just bioplastics). These materials are being designed to decay at an acceptable rate when disposed of in the natural environment. Plastic materials with a great variety and effectiveness of electrical properties are also being developed ("Plastic Manufacturing: Past, Present, and Future" 2020).

With such developments appearing on the horizon, it is hardly surprising that many experts foresee substantial growth for the plastics industry over the coming decades. One of the most commonly cited predictions comes from Canadian biologist and blogger Darrin Qualman, who argues that the amount of plastic being produced worldwide by 2050 will be four times that being produced today; a total of 1,800 Mt. Qualman's conclusion are largely supported by an important study conducted by the World Economic Forum and other prestigious organizations ("The New Plastics Economy: Rethinking the Future of Plastics" 2016; Qualman 2017).

Another way to think of the future of plastics is what your own personal experience tells you about the issue. Look back at the forms you filled out at the beginning of chapter 1 of this book. How many of the plastic items on your list could be replaced easily or even replaced at all? If your list is typical, the answer is likely to be "few" or "almost none at all." For example, you might like to replace all those plastic bottles that accumulate from daily use. What would your options be? Milk or fruit juice in glass bottles? Probably not. Cans made of tin or some other metal? Almost certainly not. Or how about personal care items, such as toothpaste or hair spray or shaving cream? You might be hard pressed to find comparable products that do *not* contain one or more kind of microplastic in the material, specifically added to make it a better product.

So part of the equation for the future of plastics is that those materials have become so much a part of our everyday lives *and* few substitutes for them are available that most of us wouldn't know how to get rid of them even if we wanted to. The conclusion one might draw from this brief review is that it looks as those plastics are here to stay, and the amount produced will almost certainly continue to increase in the future. Of course, many people are not willing to accept that conclusion. They argue that the problems caused by plastics are just so great that something(s) has to be done to reduce their presence in our lives. More on this conundrum can be found later in this chapter.

### Fossil Fuels and Plastics Production

Interestingly enough, one of the most interesting and significant aspects of any "future of plastics" scenario is the role that the world's energy companies are likely to play in that story. Fossil fuels—crude oil and natural gas in particular—are the sources of the vast majority of plastics made in the world today. Both fuels consist of complex mixtures of hydrocarbons, compounds made up of the elements carbon (C) and hydrogen (H). The hydrocarbons in these fuels are usually saturated, meaning that every carbon atom in the compound is attached to some maximum number of hydrogen atoms. Methane ($CH_4$), ethane ($C_2H_6$), and propane ($C_3H_8$) are the three simplest saturated hydrocarbons. The process from fossil fuels to plastics involves one or both of two basic steps: fractional distillation and cracking.

Fractional distillation is carried out in tall towers heated to temperatures of a few hundred degrees at their base. When crude oil is pumped into a fractionating tower, the components of which it is made boil off at different temperatures. The products of this process are, in general, known as petrochemicals. Gaseous hydrocarbons escape from the top of the tower, where they are captured for future use. Methane and ethane are the most common hydrocarbons in this fraction of the oil. Compounds with higher boiling points vaporize at lower levels of the tower. These groups of compounds are classified according to the uses for which they are largely intended: gasoline, naphtha, diesel oil, lubricating oil, fuel oil, and bitumen. (For a diagram of a fractionating tower and products obtained from it, see "Crude Oil Fractional Distillation" 2020.) The naphtha fraction contains the majority of compounds used to make plastics.

Saturated hydrocarbons are useless as raw materials for the production of plastics. They must first be converted to unsaturated hydrocarbons, compounds in which every carbon atom has fewer hydrogen atoms than it is possible to hold. The simplest unsaturated hydrocarbon is ethene (ethylene; $C_2H_4$), which contains two fewer hydrogen atoms than does ethane. The extra places on the carbon atoms in ethylene provide places at which 2 (or 3 or 4 or 50 or 100 . . .) ethylene molecules can combine with each other to form a polymer.

Cracking is one way by which saturated hydrocarbons can be converted to unsaturated hydrocarbons. For example, heating a sample of ethane gas to temperatures of 450–750°C at pressures of about 70 atmospheres (about 1,000 pounds per square inch) over a catalyst (usually a compound known as a zeolite) causes ethane molecules to "crack," giving off hydrogen atoms, and become molecules of ethylene. Cracking is often conducted in tall towers not unlike a fractionating tower. The ethylene (and propylene and other unsaturated hydrocarbons) produced in a cracking tower are captured and transferred to a plastics factory. There they are used in the manufacture of polyethylene, polypropylene, and other types of resins ("Cracking and Related Refinery Processes" 2016).

This whole process is of great significance today because of the role played by fossil fuels in the production of resins. Nowadays in the United States, about 69 percent of all crude oil consumed is used in the transportation sector, as fuel for cars, trucks, ships, trains, airplanes, and other vehicles ("Oil: Crude and Petroleum Products Explained" 2019). This use of oil is subject to widespread criticism in the world today because of the very large quantities of carbon dioxide produced and released by the transportation sector. This carbon dioxide is one of the most significant contributors to the process of climate change now being observed. Efforts are being made worldwide to reduce our dependence on oil and gas for purposes of transportation. The increasing interest in electric cars and the development of more fuel-efficient vehicles are reflections of this trend.

Energy companies that own, develop, and distribute oil and gas are well aware of this trend. Many see the 'handwriting on the wall," warning that a major use of their products—the transportation sector—is now and will continue to be a less reliable consumer of their products. They have begun to realize and act on the possibility of finding other uses for their products. Plastics production is one of the major alternatives available. Thus, for example, Royal Dutch Shell is beginning to recognize that General Motors may be a less important client for the fuels it produces in the future than is Dow Chemical, which needs oil and gas as feedstocks (raw materials) for the plastics it makes.

An especially important part of this new trend is the process of hydraulic fracturing ("fracking") in the recovery of oil and natural gas. First developed in the late 1940s, fracking has become a relatively inexpensive and highly efficient method for extracting oil and natural gas from previously inaccessible locations. The process has been responsible for a skyrocketing production of natural gas in many parts of the world, most prominently, the United States. It has resulted in the somewhat ironic consequence that more and more oil and gas are being produced just at a time when pressures to reduce use of these material are growing significantly. What to do then, energy companies ask, with all the fossil fuel resources being produced when many parts of society are saying enough? Again, the answer may be this: plastics. (A series of helpful publications about this issue is available at "Fueling Plastics: Series Examines Deep Linkages between the Fossil Fuels and Plastics Industries, and the Products They Produce" 2015.)

This new vision of a fossil fuels future is becoming a reality among energy companies at a surprisingly rapid rate, often without much public attention to the changing landscape.

One recent study revealed that 333 new projects for the use of oil and natural gas in the production of plastics in the United States alone, involving investments totaling more than $200 billion, had been initiated between 2010 and 2019. Officials of the American Chemistry Council saw this development as "an exciting milestone for American chemistry and further evidence that shale gas is a powerful engine of manufacturing growth" ("U.S. Chemical Industry Investment Linked to Shale Gas Reaches $200 Billion" 2020).

Examples of this changing role for fossil fuels in the production of plastics are widely available on the Internet. For example, two Saudi Arabian oil companies, Saudi Aramco and Sabic, are planning construction of a new facility in which 400,000 barrels of crude oil per day will be converted directly to petrochemicals. Nearly half of all the oil used in the plant will go to this purpose, compared to about 5–20 percent in traditional refineries. Even this mammoth fuels-to-plastics project pales in comparison with other such projects on the drawing boards or already in operation (Tullo 2019). In the United States, the largest such fuel-to-plastics plant designed thus far is a facility approved for construction in St. James Parish, Texas. The project, called the Sunshine Project, will cover 2,500 acres and, among other features, will be the largest emitter of greenhouse gases in the United States by 2029 (Storrow 2020).

Probably, needless to say, the growing interest in petrochemical production by energy companies has not been received with undiluted enthusiasm by all sections of society. Various individuals and organizations have pointed out that the basic purpose of such plants, an increase in the production of plastics, goes against current recommendations for solving the problem of plastics pollution. They also note that the development of new facilities is likely to increase the contribution of oil and gas use to the problem of climate change. A common concern is the increase in the number of communities exposed to the toxic products and by-products of petroleum refining. Critics have long pointed to the existence of so-called cancer alleys, where rates of cancer and other diseases appears to increase substantially with the construction and operation of petroleum plants. Indeed, they now predict that the cancer alleys previously limited largely to the middle southern states are now likely to develop in previously "clean" areas, such as Pennsylvania, where fracking is especially common (Gardiner 2019).

Concern about cancer alleys has long been a critical part of the study of "environmental justice." That term refers to the fact that major centers of air, water, and ground pollution tend to be located in low-income, nonwhite, and other minority communities. This field of study goes back at least to the early 1980s when both scholars and ordinary citizens began to realize that large solid waste facilities, industrial plants, extensive transportation centers, and other pollution-producing facilities were (and are) disproportionately located adjacent to minority communities. Residents of those communities, therefore, tend to experience unusually high levels of cancer, respiratory problems, and other health issues. Over the last four decades, substantial efforts have been devoted to learning more about this problem and finding ways of alleviating it. Sadly, those efforts have thus far achieved rather modest positive results. The trend toward the construction of even more fractionating, cracking, and fracking facilities for the production of petrochemicals is especially troubling to authorities in the field of environmental justice (Dermansky 2020).

## Disposal of Plastic Wastes

Plastic materials that have ended their useful life have one of three fates: They can be disposed of in some designated location on land, such as an open dump or landfill; they can be incinerated; or they can be recycled. The likelihood of any one of these events occurring depends on many factors, one of the most important being which country is being discussed. For example, some countries tend to recycle their plastic wastes more efficiently than do others. In Great Britain, the recycling rate for plastics today is about 30 percent; in China, about 25 percent; in the United States and most other nations of the world, about 10 percent (Geyer, Jambeck, and Law 2017, 3). The recycling rate worldwide remained at less than 2 percent until the early 1990s, and in the United States, it has never reached more than 9 percent ("National Overview: Facts and Figures on Materials, Wastes and Recycling" 2019; also see table 5.7 in chapter 5 of this book). (The topic of recycling will be discussed in more details in the following sections.)

### Dumps and Landfills

Until very recently, the vast majority of plastic waste generated worldwide ended up in some form of land disposal site, such as an open dump or a landfill. People have been using open dumps for their wastes for millennia. Objects that had to be disposed of, such as used paper, metals, cloth, food scraps, glass, and other materials, have traditionally been deposited in some designated "hole in the ground." The simplest of these is an open dump, which is nothing more than an open repository for wastes. Open dumps provide no protection for the surrounding environment, and used materials can simply leach into the soil, escape into nearby waterways, or evaporate into the atmosphere.

By the middle of the twentieth century, people began to recognize the hazards posed to the natural environment and to human health by open dumps. They began looking for more sophisticated, safer alternatives to open dumps (Ward 2011). The solution to this problem was the landfill. A landfill is an area constructed to hold waste materials for extended periods of time (many years) with barriers that prevent the material deposited in them and the substances produced by their breakdown from escaping into the surrounding environment. These barriers usually consist of a thick, strong, flexible barrier at the bottom of and on the sides of the landfill. There may also be a barrier covering the landfill, preventing the escape of gaseous materials from the site. Landfills have varying degrees of sophistication, with the most complex and efficient being labeled *sanitary landfills* (Alexander 2019).

The principle behind landfills is not so much to encourage the decay of materials deposited there as to entomb them for very long periods of time. The "very long" can vary from a few weeks for some materials, such as food scraps and paper wastes, to hundreds of years for most kinds of plastic, and even millions of years for glass (Leblanc 2019). So plastics in landfills are largely a matter of "out of sight; out of mind." If we can't see the plastic waste any longer, we don't have to worry about it.

That view is not completely correct, however. Very few, if any, landfills are perfectly secure. Materials stored within them may be able to leak through cracks in the barriers and escape into the surrounding soil, ground water, and atmosphere. And some of these materials may be detrimental to human health or to the environment itself. Research on this topic is still somewhat sparse, but some studies suggest that plastic additives may be among the substances most likely to leak from a landfill and pose problems for human health.

A *plastic additive* is a material added to a plastic to improve its durability, flexibility, color, resistance to fire, density, or other property. Data on the amount of additives used in a plastic are difficult to come by, but one estimate places that number at about 7 percent by weight of a plastic material. The special importance of additives has to do with their potential harm to human health and the environment. Generally speaking, the materials from which plastic itself is made tend to exert no effects on human health or the environment. Such is not the case with additives. Several of the hundreds of additives used in plastic production have been shown to be carcinogenic, mutagenic, teratogenic, or harmful in some other way.

One such material is bisphenol A (BPA), used to line the interior of plastic bottles. The compound has been found to produce hormonal disorders, cancer, behavior problems, cardiac issues, and other health problems in humans. Another such group of additives widely used in plastics are the dioxins, a family of compounds known to be carcinogenic and harmful to a person's testosterone level. Nearly a dozen similar additives have been found to have similar deleterious effects on human health (Alabi et al. 2019; Hahladakis et al. 2018).

Worldwide, the fraction of plastic wastes going to landfills and other repositories has continued to decrease over the years. At one point (1980), essentially all waste plastics experienced that fate. Since that time, the fraction of plastic wastes going to landfills has decreased to 93 percent in 1990, 81 percent in 2000, 70 percent in 2010, and 56 percent in 2015. This decrease has occurred, of course, because of the greater use of incineration and recycling for plastic wastes (Geyer, Jambeck, and Law 2017, Figure S7).

This pattern has been different in the United States. The fraction of plastic wastes going to landfills in 1980 was, as was the case worldwide, essentially 100 percent. By 1990, that number had been reduced to 80 percent, after which it essentially leveled off (78% in 2000, 77% in 2020, and 75% in 2015). Again, this pattern reflects the failure of the United States to make use of incineration and recycling as was in disposing of wastes ("Plastics: Material-Specific Data" 2019).

Although widely used in the United States and worldwide, landfills are not a totally risk-free method for disposing of plastic wastes. The study of the effects of plastic-based leachates (materials that have been washed out of wastes in a landfill) on human health and the environment is still in its earliest stages. However, some information has begun to accumulate. First, as noted earlier, landfills are generally not completely isolated systems. Cracks develop in plastic lining and other problems arise, allowing substances from plastic waste to escape into the surrounding ground. Most commonly, these substances are additives or monomers from which the plastics are made. At least two dozen specific toxic or otherwise hazardous compounds in these categories have been identified in these leachates. Those compounds have been shown to cause a variety of health problems, such as cancer, organ problems, and hormone disruption (Alabi et al. 2019; Hahladakis et al. 2018; Klinck and Stuart 1999; Teuten et al. 2009).

## Incineration

The concept of using incineration (burning) for the commercial disposal of solid wastes was first developed in the late nineteenth century. In 1874, British engineer Albert Freyer received a patent for a furnace in which solid wastes could be incinerated. The products of the process were a mix of gases, primarily carbon dioxide and water vapor (flue gases); slag, a stony material; and ash. The slag and ash could then be deposited into an open dump or at some other waste disposal site. Freyer called his invention a *destructor*, an appropriate name for a device designed primarily to get rid of unwanted solid wastes ("Centenary History of Waste and Waste Managers in London and South East England" 2007). Destructors immediately became popular throughout Great Britain, and more than 250 of the machines were built over the next 30 years. They eventually fell out of favor, however, because they had no system for controlling the spread of incineration products, such as ash and smoke, throughout surrounding neighborhoods ("History of Waste Management" 2017).

The idea of incinerating solid wastes remained popular, however, and inventors continue to search for ways of making them more efficient and less offensive to the natural environment and human health. By 1885, then, the first trash incinerator had been constructed at a waste disposal site on Governors Island, New York. As in Great Britain, the invention quickly became popular, and about 200 similar devices were constructed at various locations throughout the country. All forms of the early incinerator were built to accept organic materials only, such as food scraps, and not the vast majority of household and industrial wastes being produced (Walch 2002).

The incineration of plastics was largely an ignored practice worldwide until the early 1980s. According to the best available data, the fraction of plastic wastes incinerated rose from less than 3 percent in the period from 1980 to 1985, to about 10 percent in 1998, to its present level of about 25 percent. The use of incineration for plastic wastes varies considerably from country to country, with a use rate of about 40 percent of all such wastes in Europe, 30 percent in China, and about 15 percent in the United States and most of the rest of the world (Geyer, Jambeck, and Law 2017, 3).

Today, most plastic waste incinerators are so-called *waste-to-energy* plants. This name comes from the role of such plants in capturing the energy produced during the combustion of plastics and using that energy for the production of electricity or heat. The first plant for the conversion of solid wastes (not plastics) to energy was built in the United States in 1898, but such facilities did not become popular until the late 1970s. The spur for the development of waste-to-energy plants in the United States was the passage of the Public Utility Regulatory Policy Act of 1978, which required electric utilities to purchase some fraction of their electrical outlets from so-called cogeneration plants, such as waste-to-energy facilities ("Methodology for Allocating Municipal Solid Waste to Biogenic and Non-Biogenic Energy" 2007).

Several forms of waste-to-energy systems have been developed and are now in use. One of the most popular systems is a *direct thermal* process, In this process, plastic wastes are burned in a large oven at temperatures of 800–1100°C, sometimes in the presence of excess oxygen. The heat produced in this process is used to boil water, and the steam produced is used to run electrical generators. This type of direct combustion system also includes filters to remove unwanted gases produced during incineration, as well as mechanisms for collecting and isolating ash and other solid residues of the process.

At least partly because of concerns over the release of hazardous gases and solid residues during direct combustion, other systems have been developed for the incineration of plastic wastes. (For a good review of all such systems, see Antony 2017.) Most common among these are *pyrolysis* and *gasification*. Both methods are designed to convert waste plastic materials into liquid or gaseous fuels. Pyrolysis is a process in which plastic wastes are heated to a temperature of 300–900°C in an oxygen-free environment. Under these conditions, plastic materials break down to form a liquid oil as well as combustible gases and solids. Gasification is a similar process that involves heating plastic wastes in a controlled amount of oxygen. With a limited amount of oxygen, complete oxidation of the plastic cannot occur, and the plastic materials instead are converted into a *syngas* (synthetic gas) consisting primarily of hydrogen and carbon monoxide (both of which are combustible) and smaller amounts of carbon dioxide. In each of these processes, then, plastic wastes are converted to solid, liquid, or gaseous fuels that can be used to generate heat or electricity (Antelava et al. 2019).

As with recycling, the rates of incineration of plastic wastes have increased substantially since 1980, when records were first available. Those rates worldwide rose from 0 percent in that year to about 5 percent in 1990, to about 11 percent in 2000, to about 14 percent in 2010, to the current rates of about 25 percent. According to one estimate, the rate of incineration for plastics may rise to about 50 percent of all wastes by 2050 (Geyer, Jambeck, and Law 2017, Figure S7). In the United States, those rates have risen from less than 1 percent in 1980 to 17 percent in 1990, after which they have remained constant at about 15 percent of all plastic disposal methods to the present day ("Plastics: Material-Specific Data" 2019). Incineration rates differ significantly from country to country depending on a variety of political, social, environmental, and other factors. For example, the European Union currently restricts the use of landfilling for organic wastes, including plastics. To compensate for that policy, members of the European Union now incinerate about 42 percent of those wastes (Royte 2019).

Critics raise a number of objections to the use of incineration as a way of disposing of plastic wastes. They point out that the technology is often relatively expensive, usually too expensive for less wealthy countries to adopt. They also say that siting of such plants requires substantial use of land areas and is likely to be unacceptable to residents of an area used for that purpose. Incineration also has the potential to pose health issues for humans and other animals in a region around a direct combustion, pyrolysis, or gasification plant. Such plants are designed to capture hazardous materials produced during incineration, but research suggests that such systems are often not completely dependable. When those systems fail, toxic materials such as heavy metals, toxic organic pollutants, and acid gases may escape into the atmosphere, posing health threats to those in the region. (A particularly vigorous opponent of plastic waste incineration is the Gaia organization, which promotes a "no-burn" policy on the issue. See "What We Do" 2020.)

## The Oceans

The oceans constitute by far the largest and most important sink in the "plastic cycle." An *environmental sink* is a natural reservoir in which some specific element or compound is stored for relatively long periods of time. As an example, sedimentary rocks in Earth's crust are an important sink in the carbon cycle. Those rocks can hold carbon for many thousands or millions of years before the element is returned to the cycle. The oceans are a major sink for most plastics and microplastics, because once those materials reach the ocean, they stay there essentially forever. Even if they break down into their monomers (very uncommon), there is no mechanism for rebuilding the plastics from those monomers. For all types of plastics, then, the oceans are truly "the end of the line" (Watts 2019).

Plastic wastes are by no means equally distributed in the world's oceans. Geographic and topographic effects, weather, waves, and other factors lead to the accumulation of plastic wastes in some regions and low concentrations in other regions. Consider a river estuary as an example. Estuaries are bodies of water where rivers run into the ocean. The change in speed of the water, from river to ocean, often causes pollution to settle out in such regions. Also, estuaries tend to be the site of many urban areas, where the river and ocean are convenient garbage disposal sites for plastics and other wastes. Data on the amount of plastics in estuaries is sparse, but most observers acknowledge that they are a major locus for plastic pollution (Buranyi 2019). One study of plastics taken from the Pearl River estuary in Hong Kong, as an example, found a total of 154,227 plastic items, of which macroplastics constituted only about 9 percent of the total number, with polystyrene fragments making up 92 percent of the remaining plastic wastes (Fok and Cheung 2015; see also McEachern et al. 2019; most existing studies of estuarine conditions focus on microplastics).

A better studied part of the plastic wastes in the oceans is the so-called Great Pacific Garbage Patch (GPGP). The region was first discovered in August 1997 by oceanographer and boat captain Charles Moore. Moore was returning to his home in California after completing the Los Angeles-to-Hawaii Transpac sailing race. Traveling with excess fuel, Moore and his crew decided to take the "long way" home, through a patch of the ocean seldom traveled by recreational or commercial boats and ships. Along the way, they encountered an area that appeared to be flooded with solid wastes, the vast majority of which were plastic materials. Two years later, Moore returned to the region to study his discovery in more detail, finding that it contained a very high concentration of wastes spread over a large region (Moore 2014).

Today, we know that the GPGP is more complex than it had seemed to Moore in the late 1990s. In the first place, it consists of at least two distinct regions, and possibly more, in the oceans. The regions coincide with ocean systems known as *gyres*. A gyre is a large system of circulating ocean currents accompanied by significant wind currents. They are caused by Earth's revolution on its axis. Five gyres have now been identified: the North and South Pacific gyres, the North and South Atlantic gyres, and the Indian Ocean gyre. With their sweeping, vacuum-like forces, all five gyres have the potential to become "garbage patches" for waste materials. Thus far, the two that have been studied most closely are the North and South Pacific gyres, together making up the Great Pacific Garbage Patch (Bergman 2011; "Great Pacific Garbage Patch" 2020; Lebreton et al. 2018).

The size of the GPGP is currently estimated to be about 1.6 million square kilometers (620,000 square miles), about twice the size of the state of Texas or three times the area of France. In one study, the number of pieces of plastic was estimated to be 1.8 trillion particles, with a total mass overall of about 80,000 metric tonnes (90,000 short tons). Ninety-two percent of the mass of plastics found in the GPGP were macroparticles, particles with a diameter greater than 0.5 cm. But the total number of microparticles was higher, at 94 percent of all particles counted in the study ("The Great Pacific Garbage Patch" 2020). Nearly half (46%) of all plastics found in the GPGP came from fishing nets, with the majority of the remainder of the plastics coming from other fishing activities, such as ropes, oyster spacers, eel traps, crates, and baskets (Parker 2018).

Some of the most interesting research on ocean plastics has focused on the vertical distribution of such wastes. What is the relative amount of plastic, that is, at the ocean's surface, at deeper levels, and at the ocean bottoms? In one of the first studies of this kind, researchers found that the number of microplastic particles at a depth of 200–600 meters was substantially greater than that found at the surface of the same water column. That is, although a view of the pollution at the GPGP sites can be disturbing, that view may actually be missing a greater problem at greater depths. The authors of that report suggested that "one of the largest and currently underappreciated reservoirs of marine microplastics may be contained within the water column and animal communities of the deep sea" (Choy et al. 2019, 1).

That prediction has now been confirmed in a variety of studies of organisms that live on the ocean bottom. One such study, for example, found that more than 72 percent of the marine organisms living on six of the world's deepest ocean trenches contained at least one microplastic particle (Jamieson et al. 2019). And in a perhaps even more amazing discovery, a researcher who descended to a depth of 10,898 meters (nearly 7 miles) in the Mariana Trench found and photographed a plastic bag floating on the ocean bottom (Grossman 2018).

## Environmental Effects of Plastics

What effects do plastics have on the natural environment and human health? In the third decade of the twenty-first century, that question has become one of widespread, sometimes urgent, attention among researchers, environmental specialists, governmental agencies, nongovernmental organizations, and the general public. So it is perhaps somewhat surprising that little or no concern was expressed about such questions for the first half century of the "plastic era," beginning in the 1960s. Indeed, scholarly research on the topic was largely absent until well into the beginning of the twenty-first century.

Part of the reason for this fact was that plastics certainly appeared to be safe for the environment and for human health. After all, by the end of the twentieth century, they were being used in almost every imaginable aspect of human life, from plastics bottles and other containers to grocery bags, to clothing and furniture, and even to medical devices and equipment. So how could plastics pose a threat to human health or to the environment? This belief was also, hardly surprisingly, encouraged by the plastic industry itself (Lerner 2019; Root 2019).

Gradually, however, photographs of and stories about wildlife who died or suffered from exposure to waste plastics seemed to waken public consciousness to the possibility that plastics were not entirely risk free, either for animals or for humans themselves. And researchers began to explore this field as an important and potentially rich field of study (e.g., see Ruiz-Grossman and Dahlen 2017). That research was slow to develop, however, and summaries of its results well into the 2000s reported only modest results (Thompson et al. 2009).

Some of the earliest results about health risks from plastics focused on substances used as additives in their production. Many of these additives were well known and their health effects had been well studied. For example, a class of compounds widely used as plasticizers in many plastic materials is phthalates. These compounds have long been known as hormone disrupters, substances that prevent the normal sexual system from developing and functioning normally.

Perhaps the best-known instances in recent years of concern over the health effects of plastic products has involved the compound known as BPA. BPA is a raw material used in the production of polycarbonate and epoxy resins. The compound was first prepared in 1891 by Russian chemist Aleksandr Dianin. It found no practical application, however, until the 1950s, when polycarbonates and epoxy resins were first invented and developed. These products rapidly became very popular and are used today in a host of products, including plastic bottles, food storage containers (such as "Tupperware" products), baby bottles, sports equipment, CDs, DVDs, water pipes, as coatings on the inside of food and beverage cans, and in the production of thermal paper, like that used in sales receipts. By the 1970s, one manager at the Shell Chemical Company noted that products containing BPA "now serve virtually every major US industry, either directly or indirectly" (Vogel 2009). BPA is currently one of the world's most common synthetic chemicals, with an estimated annual production of about 6 million tons (Cornwall 2020).

As early as the 1930s, however, some researchers began to raise questions about possible health effects of BPA on humans. In one historic study, British biochemist Edward Charles Dodds found that BPA displays estrogen-like properties. He did not pursue the practical applications of that discovery, however, because he then found an even more powerful synthetic estrogen, diethylstilbestrol (DES), which was put to commercial use. That compound has a complex history of its own, which is not relevant to the present discussion (for more information on this topic, see Reed and Fenton 2013).

A large body of research about the effects of BPA on human health has accumulated over the past 50 years. In sum, those data seem to suggest that the levels of BPA found in the body at which adverse health effects begin to occur is significantly less than the amount of BPA found in consumer products. Until recently, therefore, most governmental agencies worldwide have concluded that BPA poses no serious hazard to human health. As a consequence, very few regulations on its use have been issued. In 1976, for example, BPA was listed as one of about 62,000 chemicals designated as "presumed safe" by the Environmental Protection Agency as part of the newly adopted Toxic Substances Control Act (Houlihan, Lunder, and Jacob 2011).

Even today, the Food and Drug Administration is very clear about its position on products containing BPA. "FDA has performed extensive research reviewed hundreds of studies about BPA's safety," it says. "We reassure consumers," it goes on, "that current approved uses of BPA in food containers and packaging are safe. FDA continues to monitor the scientific literature for new research that helps enhance our understanding of BPA, and will consider new data as it continues to ensure the safe use of BPA in food packaging" ("Questions & Answers on Bisphenol A [BPA] Use in Food Contact Applications" 2020).

Official statements such as these still do not convince many consumers about the possible risks posed by BPA to human health. For example, one recent review of the chemical suggested that it might be responsible for infertility in men and women, heart disease, type 2 diabetes, obesity, asthma, liver disease, and thyroid, immune system, and brain function (Petre 2020). It appears likely that the debate over the health effects of BPA may continue for some time into the future.

The one important exception to this conclusion occurred in 2012, when the FDA withdrew its approval for the use of products containing BPA in certain specific consumer items: baby bottles, sippy cups, and infant formula packaging. The interesting point about this change in policy is that it was based not on safety but on "market abandonment." That is, the industry itself had developed safe alternatives to BPA in its production of the named products. So no approval by FDA was any longer needed. The bottom line was that FDA still believed that *all* products containing BPA were safe for consumer use. (Interestingly enough, later research has shown that even "BPA-free" products may pose a threat to human health, although what those threats are and how serious they might be has not been determined. See, e.g., Horan et al. 2018 and Mao et al. 2020.)

Today, the store of knowledge about the effects of plastic pollution has greatly increased over that from previous decades. The following sections focus on our evolving understanding of the harms caused by macroplastics and microplastics on the environment and human health.

## Soil

Research on microplastics in soils has progressed slowly, especially in comparison to aquatic studies. One reason for this difference is that plastic pollution in lakes, rivers, and the oceans is often relatively visible, with its effects on animals especially noticeable. Changes that may be taking place in the ground and the effects on living organisms there may be more difficult to see and measure. Nonetheless, the significance of microplastics on terrestrial environments may be as much as 23 times as great as that for aquatic environments (Horton et al. 2017). Today, however, a growing body of knowledge provides significant information about the amount of microplastics found in soils at many different locations, from agricultural sites to urban settings, to remote Swiss floodplains (Scheurer and Bigalke 2018). (A discussion of the role of nanoplastics in the soil is not included here because the size of nanoparticles is below the level at which meaningful studies can currently be conducted. See Rillig et al. 2019.)

The microplastics found in soil generally come from one of three major sources, the most important of which is sludge. The sludge produced in wastewater treatment plants is commonly sold to agricultural operations, where it is spread on fields as a fertilizer. According to one estimate, a single kilogram of sludge may contain anywhere between 1,000 and 24,000 pieces of microplastics. These particles become part of the native soil and may remain there for many years or decades. There they may be absorbed by living organisms in the soil or become part of the human diet (Bläsing and Amelung 2018). Another important source of microplastics in soil is the wastewater itself released from treatment plants. This water may also carry within it significant amounts of microplastics that would have passed through the treatment process. Water that has been used for agricultural irrigation operations has been found to contain 1,000–627,000 plastic items per cubic meter for untreated wastewater and 0–125,000 plastic pieces for cubic meter for treated wastewater. Interestingly enough, even supposedly "clean" water from rivers and lakes used for irrigation may have significant levels of microplastic particles, 0.82–4.42 plastic items per cubic meter for lakes and 0–13,751 items per square kilometer for river water (Bläsing and Amelung 2018).

Another source of microplastic pollution in the soil is plastic mulching. Plastic mulching involves the use of thin sheets of plastics placed between rows of crops to help the soil retain moisture and reduce the growth of weeds. It is a simple, often less expensive alternative to traditional mulching with organic materials. It stands to reason that some amount of plastic film from this process will break off and escape into the soil. By some estimates, anywhere from 60–500 kilograms per hectare (50–450 pounds per acre) of plastic material may remain in the soil after the removal of mulching films. As with other cases, this material may remain in the soil for substantial periods of time and become incorporated within or on to crop plants growing in the area (Hayes 2019).

The effects of microplastics on the soil seem to be complex and contradictory. The two factors that appear to effect these results most significantly are plant species and plastic type. In their recent review of the subject, Matthias Rilling and his colleagues at the Institute for Biology and the Free University in Berlin summarized these effects. They concluded microplastic fibers were likely to produce the largest effect on plant growth in a positive direction (i.e., the more the microplastic fibers, the greater the plant growth). Other types of microplastics were likely to result in minor effects on plant growth (beads and fragments) or in intermediate negative effects on plant growth (films and biodegradable microplastics) (Rillig et al. 2019, Table 1, 1067). These results appear to occur because of physical changes that microplastics make in the soil in which plants are grown. They may, for example, increase the bulkiness of soils, making root growth more difficult, immobilize nutrients needed by a plant for its growth, or exert toxic effects directly on a plant (Rillig et al. 2019; Machado et al. 2018).

Thus far, little research has been directed at detailed studies of the effects of microplastics on organisms that live in the soil and that are part of the normal biota of that part of the environment. One interesting study has found that earthworms that live in soil contaminated with microplastics tend not to grow normally but to lose weight when compared to control worms living in microplastic-free soils. The authors of this report speculate that this effect may occur because of the worm's inclusion of nonnutritious plastics in their diets, thus limiting or reducing the nutrients needed to achieve normal growth and development (Boots, Russell, and Green 2019).

### Entanglement

Most of the harm caused by macroplastics to animals comes in one of three forms: entanglement, ingestion, or interaction (Law 2017). Entanglement is the process by which an animal becomes trapped to one extent or another in some plastic material. That material may prevent the animal from capturing and swallowing food, from growing normally, and from being able to swim or otherwise move around in the environment, or to become disabled from physical injuries ("Entanglement: Entanglement of Marine Species in Marine Debris with an Emphasis on Species in the United States" 2014). Many types of plastics are responsible for entanglement events. One of the most commonly seen cases in articles on the topic involves an animal having been trapped by the plastic rings that make up a six-pack bottle holder (see, e.g., Messenger 2015).

The most common cause of entanglement, however, is fishing gear, such as fishing lines, nets, traps, and other equipment. The most frequent culprit within this category is so-called ghost fishing equipment. The term *ghost fishing* refers to the reality that large amounts of legal common fishing gear are intentionally discarded into the oceans or along beaches or are accidentally lost by boats at sea. These materials often float on the ocean's surface or sink to the ocean floor, where they continue to catch fish, turtles, sharks, and other organisms ("Fishing's Phantom Menace: How Ghost Fishing Gear Is Endangering Our Sea Life" 2014). One study of entangled sharks and rays found that nearly three-quarters of entanglement cases (74%) were caused by ghost fishing gear. Another 11 percent resulted from contact with polypropylene strapping bands and the remaining cases from interaction with circular plastic debris, polythene bags, and rubber tires (Parton, Galloway, and Godley 2019).

Entanglement has become a major issue of concern in the damage caused by plastics to animals in the oceans and, to a lesser extent, terrestrial animals. Research suggests that significant fractions of many marine species are affected by entanglement, including 100 percent of all seven species of marine turtles, 67 percent of all seal species, 31 percent of all whale species, and 25 percent of all seabird species (Entanglement n.d.). A 2014 study found that entanglement has become a matter of environment concern in the United States for 111 different species, ranging from crabs, clams, seals, dolphins, and whales to more than 40 species of sea birds and to more than a dozen species of sharks and sea turtles ("Entanglement: Entanglement of Marine Species in Marine Debris with an Emphasis on Species in the United States" 2014, Appendix). Some of the effects that have been attributed to entanglement in plastics in the oceans include body abrasions, damage to the skin, various types of wounds, changes in population size and structure, and death (Law 2017, Table 1, 117).

One of the most significant programs for dealing with entanglement of marine animals is the Marine Animal Entanglement Response (MAER) program of the Center for Coastal Studies. Since its founding in 1984, MAER has responded to more than 200 notices of marine animal entanglement, using techniques especially developed for dealing with this problem. More information on the program is available at https://coastalstudies.org/rescue/.

## Ingestion

Another important plastic-related issue for marine organisms is ingestion of macroplastics and microplastics. Given the abundance of both macroplastics and microplastics in the world's oceans, it would be surprising if ingestion by a whole range of species were not the case. Indeed, one important review of the topic found that anywhere from a few percent to nearly 100 percent of most species studied were found to have ingested some quantity of plastics (Thiel et al. 2018). Interestingly enough, ingestion of plastic materials can be observed even in the simplest of organisms, such as phytoplankton and zooplankton. This process often has no observable effect on the organism's functioning. But deleterious effects have also been observed. For example, ingestion of microplastics has been found to inhibit photosynthesis, decrease feeding patterns, inhibit growth, and cause death among both phytoplankton and zooplankton. Similar results have been found in most classes of benthic invertebrate, as well as almost all species of marine vertebrates (Lusher 2015).

The magnitude of this problem has now become familiar both to members of the research community and to the general public. Numerous photos of animals that have ingested seemingly impossible volumes of macroplastics have appeared in the popular media. In one example, a sperm whale weighing 20 tons washed up on a Scottish beach in December 2019. During an autopsy of the animal, its stomach was cut open to reveal a ball of mixed plastic objects weighing about 220 pounds (Collman 2019; note: the images in this article may be difficult to look at).

The pervasiveness of macroplastics and microplastics in marine organism diets has become apparent from a variety of studies. For example, a recent study has found that fish even at the earliest stages of their life may begin to ingest plastic materials in much the same way they ingest objects in their natural diet. Researchers dissected 658 baby fish off the west coast of the Hawaiian Islands and found that 42 contained some form of plastic particles. Authors of the report noted that their findings did not bode well either for the possible survivors of fish themselves or for the fishing industry that is also under pressure from other anthropogenic factors in today's world (Gover et al. 2019).

One obvious question associated with this phenomenon is why marine organisms ingest plastics in the first place. This question has received considerable attention by researchers, and some clues are now available to possible answers. In general, ingestion by marine animals occurs largely because of the fact that such animals do not necessarily find and select foods in the same way humans do. Of course, some similarities do exist. It seems likely that many of the plastic-as-food choices that animals make are based on the fact that a piece of plastic looks like the kind of food with which the animal is familiar. For example, one study of green turtles washed up on the shores of Cyprus found an abundance of macroplastic that looked very much like the seagrasses that make up the turtle's normal diet (Duncan et al. 2019).

At other times, animals use other senses, the sense of smell being perhaps the most common. Many species of animals rely far more on smell than do humans, although smell is something of a factor for humans too. For examples, some studies show that anchovies choose plastics in their diet because the plastic smells like their normal food, krill (Savoca et al. 2017). Seabird species also demonstrate a similar pattern. One study showed that seabirds tend to be attracted to the compound dimethyl sulfide, produced during the decay of marine algae. The compound is also produced by the breakdown of some types of plastics, such that seabirds may not be able to recognize the difference between natural products in their diet (algae) and plastics (Savoca et al. 2016).

Animals also can use other methods for locating food that are essentially unavailable to humans. Perhaps the best known of these methods is echolocation. Most species of whales and dolphins use this system for location of food. They send out sonar waves to their surrounding environment and then identify sound patterns that are characteristic of their prey. In some cases, macroplastics may produce echolocation patterns similar to those of the prey, and predators ingest them as a possible source of food (Daly 2019).

Harmful effects that can occur among marine animals as a result of ingestion include intestinal blockage, perforation of the intestinal lining, modifications of biochemical systems, gastric rupture, ulcerative lesions, reduced organ size, altered population sizes and structures, and death (Law 2017, Table 1, 117–118).

### Interaction

The term *interaction* is used in this context to refer to events in which an animal comes into contact with a piece of plastic, without there also being entanglement or ingestion. Some examples include the collision between invertebrates and fishing gear, such as fishing lines and lobster traps; between sea turtles and medical wastes on or in the water; and between marine insects and microplastics. Such interactions can result in problems such as simple abrasions, changes in population size of the species, suffocation, and death (Law 2017, Table 1, 117–118). Events of this type appear to be less common and, in general, less harmful to organisms than do entanglement and ingestion (also see Rochman et al. 2016).

## Human Health

Hundreds of studies have been conducted in the past two decades on the effects of macroplastics on human health. The vast majority of these studies focus on one of three aspects of the problem: the impact of some specific component of plastics, such as BPA or phthalates, as discussed earlier; the mechanism by which a plastic component enters the human body and makes some specific change; or the specific disorders that can result from exposure to a macroplastic or macroplastic component (see, as only one example, Kumar 2018).

In 2019, a consortium of eight nongovernmental agencies issued a report outlining an alternative view of the question of how plastics affect human health. That report suggested that existing data are of limited value because they have been collected on a "piecemeal" basis focusing primarily on plastic use and disposal. It recommends a new "lifestyle" approach, in which the hazards posed by plastics should be considered at every stage of their existence, "from wellhead to refinery, from store shelves to human bodies, and from disposal to ongoing impacts as air pollutants and ocean plastic" (Azoulay et al. 2019, 1). Authors of the report remind readers that health effects occur at every stage of the production and use of plastics, from the extraction and transportation of feedstocks to the refining and actual production of plastics, to the manufacture of plastic products and their packaging, to the disposal of plastic wastes, and to the formation of microplastics and their own specific health effects. Some of the effects that have thus far been found include the release of more than 170 toxic and carcinogenic chemicals during the refining of fossil fuels used in the production of plastics; the exposure of humans to gaseous, liquid, and solid hazardous chemicals during the production of plastics; the many effects resulting from physical contact of humans with consumer products containing plastics and plastic additives; the exposure of humans to plastic wastes at various steps in the disposal process; and the ingestion of macroplastic and microplastic particles simply during the everyday processes of respiration and ingestion (Azoulay et al. 2019, 8).

In addition to this excellent review of existing information on the health effects of macroplastics, the 2019 report summarizes much of what is known about the effects of microplastics, along with the kind of research that is still needed in that area. One line of that research suggests that microplastics may enter the body through any one (or all) of three main pathways: from the air through respiration and from water and food that is ingested. A recent study conducted at the Medical University of Vienna and the Environmental Agency of Austria confirmed the widespread exposure of humans to microplastics. That study looked for the presence of microplastic particles and resin monomers in the stools of subjects from eight countries (Austria, Finland, Italy, Japan, the Netherlands, Poland, Russia, and the United Kingdom). It found residues of microplastics and nine different kinds of resins in all subjects (Schwabi et al. 2019).

Studies of the effects of airborne microplastics on human health are sparse. One review of the topic found only 13 studies have been conducted and reported as of 2019. Those studies have found no identifiable health issues associated with inhalation of microparticles, although a variety of such effects have been hypothesized. Those effects include "localized inflammation and cancer due to responses by the immune cells, especially in individuals with compromised metabolism and poor clearance mechanisms" (Enyou et al. 2019).

Several studies have found that microplastics are common in drinking water. In one pair of studies, researchers found significant amounts of microplastics in both community drinking water systems and bottled water. In the former case, data came from 159 tap water supplies from 14 countries, of which half were developed and half developing countries. In eighty-one of the samples, at least some microplastics were found, ranging as high as 61 microparticles per liter. The average concentration of microparticles was 5.45 particles per liter, with the highest level in water from the United States, at 9.24 microparticles per liter (Kosuth, Mason, and Wattenberg 2018; Mason, Welch, and Neratko 2018; for a review of 50 studies on microplastics in drinking water, see Koelmans et al. 2019). In a comprehensive review of research on microplastics in drinking water, the World Health Organization concluded that "there are no studies on the impacts of ingested microplastics on human health and there are only a limited number of animal studies of questionable reliability and relevance. Some data suggest a very limited uptake and impact of microplastics <50 μm in laboratory animals at high concentrations, but the relevance to humans is unknown. These studies require confirmation under realistic exposure conditions before firm conclusions can be drawn" ("Microplastics in Drinking Water" 2019, 64).

Concerns about possible dangers to human health associated with ingestion of microplastics in foods and other solid substances are widespread on the Internet and other public media. Most such concerns are based on research on the effects of microplastics on experimental animals and on studies of some limited aspect of the microplastic-to-risk process. Fish and shellfish have been a particularly popular subject for such research, primarily because such animals play a critical part in the diet of millions of humans around the world. A common pattern is to expose various species of organisms to certain levels of microplastics and determine any physical, biochemical, physiological, reproductive, or other bodily change that result from such an exposure. It is not unusual for researchers to find changes in fecundity, growth patterns, organ function, or other changes in such cases. Those data may then sometimes be used to extrapolate health effects on humans ("Effects of Microplastics on Fish and Invertebrates" 2018).

One problem with this approach is that in many cases a host of possible factors may be involved in the determination of harm to an organism. It may be difficult, therefore, to generalize from any one study about microplastics and fish to broader conclusions posed by exposure to microplastics. Metastudies that have attempted to summarize the status of microplastic to fish and similar research have generally found that there is little basis for predicting human health effects from studies of lower animals. As one highly respected report has concluded:

> Little is known with respect to the ecological and human health risks of NMPs, [nano- and microplastics] and what is known is surrounded by considerable uncertainty (Section 2.6). For microplastics, from the current evidence, the working group has formulated three conclusions with respect to ecological risks: one concerning present local risks, one concerning present widespread risks and one concerning the likeliness of ecological risks in the future.

Respectively, these conclusions are:

- There may at present be at least some locations where the predicted or measured environmental concentration exceeds the predicted no effect level. . . .
- This means there may be some selected specific locations where there is a risk.
- Given the current generally large differences between known measured environmental concentrations (MEC) and predicted no effect levels (PNEC), it is more likely than not that ecological risks of microplastics are rare. . . . This means that the occurrence of locations with risks is rare.
- If microplastic emissions to the environment remain the same, the ecological risks of microplastics may be widespread within a century. . . . This means that, if NMPs continue to be emitted or formed from larger plastic debris as they do now, without any restriction in the future, that there could be widespread future risks

in most locations. ("A Scientific Perspective on Micro-plastics in Nature and Society" 2019)

The preceding final section represents a common view among many experts in the field, namely that direct effects between microplastic exposure and specific human disorders may still be lacking. But adequate data exist to cause legitimate concerns about future trends in that pattern. Another important summary of the microplastics-to-human health pathway drew a somewhat different conclusion from that stated earlier. It noted that many existing studies frequently cited in the literature do not provide sufficient evidence to draw specific conclusions about the health effects of microplastics. Still, there is good reason to think that microparticles may be associated with a wide range of human health conditions, including renal, cardiovascular, gastrointes-tinal, neurological, reproductive, and respiratory systems such as cancers, diabetes, and developmental toxicity (Azoulay et al. 2019; Public Health Standing Committee 2015).

Microplastics compromise human health through a variety of mechanisms. In some cases, a microplastic itself may be, or may contain, an element or compound known to be toxic to humans. The degree of toxicity has been shown to be the con-sequence of many factors relating to the microparticle, such as size, shape, electrical charge on the exterior, and solubility of the material. Microplastics may also cause bodily harm simply because of physical damage to surrounding tissue. Microparticles shed from surgical implants are one source of this effect. The wear and tear of such contact may result in inflammation and infection of adjacent tissues. Some harm is likely to occur also from toxic particles adsorbed on the surface of microparticles and transmitted to the digestive sys-tem. Some research also suggests that the addition of micro-plastics to the human digestive system may cause changes to the natural biome of bacteria, with potential damage to the digestive system (for a summary of these effects, see Wright and Kelly 2017).

## Solutions

Few people or organizations today will argue that plastic pollution is a serious problem about which the world needs to be concerned or, at the very least, that accumulations of plastics, microplastics, and nanoplastics in the environment pose a threat to the world's future. Most possible solutions to this problem fall into one of two general categories: (1) finding ways of collecting and recycling used plastic and (2) reducing the amount of plastic wastes produced in the first place.

## Recycling

The recycling of plastics is usually a complex and expensive operation. It may consist of any one or more of three steps: primary, secondary, and tertiary recycling. (A quaternary form of recycling is also used, involving the conversion of wastes to energy. The terminology of plastic recycling is a bit confusing. See Hopewell, Dvorak, and Kosior 2009, Table 2.) The steps used depend on a number of factors, including the type of plastic involved, the kinds of plastics present in a mixture, and the kinds and amounts of impurities present in the plastic. The following discussion, therefore, can focus only on some general principles in the recycling process. For more technical details, see, for example, Al-Salem, Lettieri, and Baeyens 2009.

Most plastic wastes must pass through a series of preparatory steps before recycling can actually occur. They must, of course, first be collected by consumers and set aside for pickup by some type of waste management program. They are then combined by some type of baling process for shipment to a recycling plant. At the plant, the wastes must be washed to remove any foreign materials that might interfere with the actual recycling process. They are then shredded into finer particles that can be used in the manufacture of new plastic materials. At some point in this process, or as a result of the process, various types of plastics incompatible with each other must be sorted out from each other (Jeavans 2008; see graphic).

In primary recycling, this process may be adequate for the production of resin pellets that can be used for the production of new plastics of the same type contained in the wastes. For this reason, primary recycling is also called *re-extrusion*. Extrusion is the process by which liquid or solid pellets are passed through a die of desired shape to produce a material with the correct shape and dimensions. Primary recycling is used most commonly with relatively large pieces of plastic that can easily be separated by type, such as scraps from the manufacturing process, and large used pieces, such as automobile doors and instrument panels (Al-Salem, Lettieri, and Baeyens 2010). Primary recycling can be carried out with relatively modest preparation of the waste product and, therefore, tends to be a desirable choice for recycling programs.

Primary recycling is considered to be a form of *closed-loop recycling*. In closed-loop recycling, the material used in the manufacture of a plastic in the first place is recovered from the waste material and then reused as a raw material for the original plastic. In other words, the plastic material can be recovered and reused over and over again, in a closed cycle (Roegner 2019; for a helpful schematic, see "Plastics Recycling: Overview and Challenges" 2020). Closed-loop recycling is a highly desirable option to open-loop recycling. In open-loop recycling, the products recovered from the recycling process are different from those used in the original production of the plastic material. One example of open-loop recycling involves the conversion of plastic bottles into a new type of plastic material that can be used as a fabric for fleece jackets or sleeping bags. The process is called "open" because the product obtained from recycling (fleece, in this case) cannot itself by recycled; it must be thrown out, thereby leaving the plastic cycle (Kaye 2013).

Secondary recycling is also commonly known as *mechanical recycling* because it makes use entirely of physical methods of recycling. The process is relatively straightforward. Plastics are sorted by type, of which only polyethylene, polypropylene, and polyethylene terephthalate can be used in mechanical recycling.

The different types of useable plastics are then separated from each, either by hand or by some mechanical means. They are then washed and ground into flakes that, in turn, are converted to resin pellets. These pellets can then be used as raw material for the production of a new plastic material (Al-Salem, Lettieri, and Baeyens 2009, 2628–2631; Delva et al. 2019).

Tertiary recycling is also referred to as *chemical recycling* or *advanced recycling*. As the name suggests, chemical recycling involves the treatment of plastic wastes with chemical reactions that break the polymers of which they are made down into their constituent monomers. In the simplest possible sense, the procedure is essentially the reverse of the process by which polymeric plastics are made in the first place:

Original production: monomers → polymers
Tertiary recycling: polymers → monomers → polymers

Although this process may sound like a relatively simple approach to the recycling of plastic wastes, it involves several challenging technical problems. Currently, it is used most commonly with polyethylene terephthalate (PET) and certain kinds of polyamides, especially nylon 6 and nylon 66. Extending this process to other types of plastics is currently an active field of research (Phipps 2019; "What Is Advanced Recycling?" 2020).

The story of plastic recycling differs significantly from country to country. In general, a rule of thumb is that the amount of recycling in a developed country depends on regulations at the state and local levels. In regions where recycling is seen as a positive benefit, those governmental units are likely to adopt regulations that require or encourage recycling. That trend exists in the European Union where about 30 percent of all plastic waste is recycled. In developing countries, the determining factor in plastic recycling appears to be the needs of the industry in a specific nation. If the recycling of plastic wastes can be shown to have some economic value, residents will adopt the practice and become an essential part of the business of recycling. Local rules and regulations with regard to recycling are largely absent from such communities. Examples of nations that fit this model include China, Indonesia, Sri Lanka, Philippines, Vietnam, and Malaysia (d'Ambrières 2019).

For the most part, recycling of plastics takes place in or near the country of origin for the wastes. Since the end of the twentieth century, however, a new pattern has also developed in which developed countries tend to export significant fractions of their plastic wastes to developing countries for their processing and reuse. In the first six months of 2017, for example, the United States shipped 379,000 metric tons of plastic wastes to China and 257,000 metric tons to Hong Kong, by far the two countries accepting the greatest quantity of wastes. The pattern changed drastically in early 2018, after the Chinese government would no longer accept U.S. plastic wastes. The corresponding numbers for the first six months of 2018, then, were 30,000 metric tons to China and 60,000 metric tons to Hong Kong ("China Won't Accept U.S. Plastic Waste. Now What?" 2019).

For a short period of time, it appeared that other Asian countries might be willing to pick up the slack in the U.S. plastic waste equation. But nations such as Malaysia, Thailand, and Vietnam quickly followed China's lead and decided not to accept any further wastes from the United States. Today, worldwide the nations most open to those wastes are developing nations such as Senegal, Philippines, Bangladesh, Ghana in Africa and Southeast Asia and Ecuador in South America (McCormick et al. 2019).

One troubling fact about this international transmission of plastic wastes is that large quantities of that material are not actually recycled in the accepting countries. Instead, significant amounts of the wastes may actually be dumped into landfills or, more commonly, into the oceans. According to one study, between 280,000 and 730,000 metric tons of Vietnam's 1.83 million metric tons of plastic waste ended up in the oceans. Comparable numbers for Bangladesh show that between 120,000 and 310,000 metric tons of its 790,000 metric tons of waste were dumped in the oceans, while, in comparison, between 40,000 and 110,000 metric tons of its total of 280,000 metric tons ended up in the seas (d'Ambrières 2019). These data suggest that important questions remain as to where plastic wastes are to be sent (or kept) for recycling treatment and how the process of exchange between nations can be developed to the greatest benefit of the world's environment.

## Biodegradable Plastics

Researchers have long been intrigued by the possibility of making biodegradable plastics, also known as bioplastics. Such products are formally defined by the internationally recognized institute for the definition of technical terms, ASTM International (formerly known as American Society for Testing and Materials), as "products (including packaging) using plastic films or sheets [that] will compost satisfactorily, including biodegrading at a rate comparable to known compostable materials" ("Biodegradable Plastics" n.d.). Such materials would, of course, be a revolutionary invention for the plastics industry. Products made of biodegradable plastic could be deposited in dumps or landfills, where they would deteriorate in essentially the same way as garbage and other organic wastes.

The earliest research on biodegradable plastics dates to the late nineteenth century with the demand for a new type of chalkboard that was white rather than black. In response to that request, German industrialist Wilhelm Krische found a way of making a plastic-like material by mixing casein, a protein derived from milk, with formaldehyde. The product of that reaction was a tough white material that could be cut, drilled, embossed, dyed, and processed in other ways. Unfortunately, it did not meet the demands of potential white chalkboard customers. Krische did, however, find other uses for this product, which he called *galalith* (from the Greek *gale* for "milk" and *lith* for "stone"). It turned out to be an ideal material for the manufacture of jewelry, a purpose for which it is still used today (Trimborn 2020).

The notion was that a plastic made from some natural material was likely to degrade in much the same way as the material itself. Thus, over the years, researchers have experimented with a host of natural materials for the production of bioplastics, including sugarcane; corn; rice; switch grass; cellulose; vegetable oil; industrial and agricultural wastes; wastewater and activated sludge, milk whey, lignin, and other waste products from the pulp and paper industry; and bacteria active in the process of biodegradation of organic substances (Pathak, Sneha, and Matthew 2014; Ralston and Osswald 2008).

As of 2020, bioplastics still constituted a very small portion of the plastic manufacturing industry. Less than 1 percent of all plastics made worldwide were classified as biodegradable. But market analysts predicted a rapid rise in bioplastic production over the next five years, growing from a valuation of about $21 million in 2017 to a projected $69 million in 2024, a growth rate of about 20 percent (Barrett 2019). The most promising materials for future biodegradable plastics appear to be polylactic acid (PLA), polyhydroxybutyrate (PHB), polybutylene succinate (PBS), hemp, and lignin. Both of the latter two substances are derived from natural products that are easily and inexpensively obtained. Overall, progress in the field is handicapped, however, by the relatively high cost of development and production and the challenges faced in making products that are reliably biodegradable (Farah 2020; a good collection on various aspects of bioplastics can be found at Articles on Bioplastics 2020).

## Zero Waste

One of the most significant events in the recent history of plastics has been the rise of the Zero Waste movement. The movement had its origins in the early 2000s when American environmentalist Richard Anthony was asked to review papers submitted at a conference sponsored by EMPA, the Swiss Federal Laboratories for Material Science and Technology on plastic waste management. He was disturbed to find that the majority of papers dealt with incineration of wastes and that relatively few considered the possibility of recycling or other methods for reducing wastes. He was invited by the chair of that meeting to assemble a group of experts who could present a more extended and cohesive explanation of that view of the treatment of plastic wastes. Out of that effort grew a movement that came to be known as Zero Waste. Zero waste is based on the principle that post-consumer plastic wastes should not be directed to landfills, incineration, or other "end-use" destinations. Instead, they should remain in the overall plastic cycle, which involves recycling as a principle element (Anthony 2013).

The concept behind zero waste illustrates a sharp contrast between competing views of waste management in general. The traditional model is a linear approach, passing from extraction of resources to production to consumer use to disposal. This model has sometimes been referred to as a "take, make, consume, throw away" approach to materials such as plastic wastes. By contrast, the zero waste model can be classified as one of "take, make, consume, reuse." Zero waste aims to take actions at every step of this process to reduce or eliminate waste. For example, programs are now being developed to reduce wastes at the very first stage of this process, the production of reusable materials. As one example, many companies are now beginning to rethink the use of plastic for drink containers. Products such as milk, bottled water, and juice and soda drinks were once readily available in glass containers. Over time, glass almost entirely disappeared as a container for such liquids. Today, many companies that sell liquid products are considering returning to the older model, using glass or metal containers for their products. Both glass and metal can be reused, making them legitimate substances for use in a zero waste economy (Sipper 2019; for one example of this change, see Parker 2020).

Another major theme in zero waste is interruption of the process involving "end of use" to "disposal." Advocates of the philosophy have recommended many steps that individuals can take to prevent the release of used plastics to landfills, incinerators, or other disposal systems. For example, one website has listed "101 easy eco-friendly" steps one can take to achieve this goal. Those steps include avoiding the use of plastic straws, donating used clothing to secondhand stores, washing clothes only when they get dirty and not just on a routine basis, using home canning as a way of preserving foods, using natural methods for pest control, making and using homemade mouthwash, ordering coffee and other drinks in glass or other reusable materials, and wrapping presents in newsprint (or not at all) ("101 Easy Eco Friendly, Zero Waste Tips" 2017).

Several governmental agencies have also weighed in on this problem, acknowledging at least in part that the plastic waste problem may not be solvable just by informed consumer actions. As noted in chapter 1, a handful of national governments, many state governments, and even more municipalities have adopted legislation decided to eliminate the flow of used plastics to disposal sites. As of early 2020, eight states (California, Connecticut, Delaware, Hawaii, Maine, New York, Oregon, and Vermont) had banned the use of single plastic bags entirely. Another 15 states had adopted other limitations on plastic bag use, usually a small fee for the use of each bag. In addition, more than 400 local cities, towns, and counties had enacted their own bans on the use of plastic bags, straws, and other objects ("Find Bag Legislation in Your Area" 2019; Nace 2018).

**Circular Economy**

A concept similar to that of zero waste is *circular economy*. The term has been defined as a system whose purpose it is to "redefine growth, focusing on positive society wide benefits." The three critical components of the system is to "design out waste and pollution; Keep products and materials in use; and Regenerate natural systems" ("What Is a Circular Economy?" 2017). That description sounds similar to that of a zero waste program, and the two terms are sometimes used interchangeably. One major difference that has been identified between the two schemes is that zero waste ideas and programs are usually directed primarily at consumers, as a way of directing their activities to more responsible treatment of plastic materials. By contrast, the concept of a circular economy is generally aimed more specifically at manufacturers and businesses that make plastics and products made of plastics (Daigneau 2017).

The concept of a circular economy has existed among professional economists since at least the eighteenth century. Its modern expression can be traced to an important paper written in 1976 by economists Walter Stahel and Genevieve Reday, "The Potential for Substituting Manpower for Energy" (Cardoso 2018). Over time, that concept grew in importance with increasing concerns about the growth in plastic wastes and the failure of existing systems to deal with that problem. Today, the concept is perhaps advanced most strongly by the Ellen MacArthur Foundation. A detailed description of the foundation's program is available on its website at https://www.ellenmacarthurfoundation.org/. An excellent diagram illustrating the elements of a circular economy can be found at the same website, at https://www.ellenmacarthurfoundation.org/circular economy/concept/infographic.

## Public Opinion and Legislation

How important are plastic and microplastic issues to the general public and policymakers? The data presented in this chapter suggests that both groups might be concerned about the role that these substances might play in our everyday lives. It's obvious that plastics and microplastics are now ubiquitous the world around us. There is also good cause to worry about the demonstrated and possible effects of these materials on the natural environment, on terrestrial and aquatic organisms, and on human health. So what evidence is there that people are (1) aware of issues surrounding the use and disposal of plastic wastes, (2) worried about the possible consequences of these products, and (3) willing to take some type of action to deal with this problem?

On the one hand, there is almost no evidence that people anywhere in the world are particularly worried about plastics and microplastics in their lives. For example, when public opinion pollsters ask people about the problems of most concern in their lives, they almost never, on their own volition, mention plastic-related topics. In its most recent poll on issues of greatest concern to Americans, for example, Gallup Poll researchers found that people offered 47 items, ranging from race relations and crime and violence to welfare, drugs, and abortion, to terrorism and school shootings ("Most Important Problem" 2020). Any reference to plastics did not occur anywhere in the poll. The closest category on the list was environment/pollution/climate change, a category mentioned by 5 percent of respondents as "the most important" problem facing America today. Studies from other geographical areas (Australia, United Kingdom, State of Rhode Island) have provided more detailed information about public opinions about plastic issues. But most such studies have involved presenting respondents with specific questions about plastics, rather than offering free-form settings in which people can offer their own opinions (see, e.g., Pereira 2019; "Public Concern about Plastic and Packaging Waste Is Not Backed Up by Willingness to Act" 2018 [Great Britain]; and Sarena-Dilkes et al. 2019 [Australia]).

The U.S. Congress has not been particularly active in dealing with the problems of plastics pollution nationwide. Most legislations have involved pollution of the oceans or inland waterways by plastics. A handful of early acts were passed to deal with the problems of ocean pollution problems in general, all of which dealt with plastics to one degree or another. One of the first of these acts was the Marine Plastic Pollution Research and Control Act of 1987. That act was adopted as an amendment to the Act to Prevent Pollution from Ships of 1980. It prohibited the disposal of plastic wastes (and other types of wastes) within a 200-mile zone from the shore. Certain regulations were also created to prevent the pollution of facilities close to the shore by plastics ("Laws that Protect Our Oceans" 2018).

A year later, the Congress passed an even more consequential piece of legislation, the Marine Protection, Research, and Sanctuaries Act (MPRSA) (16 USC § 1431 et seq. and 33 USC § 1401 et seq.). That act had three primary purposes, prohibition of three major components: (1) prohibition of transporting wastes from the United States for the purpose of ocean dumping; (2) prohibition of transporting wastes from any-where in the world on U.S.-flagged ships for the purpose of ocean dumping; and (3) prohibition of the transport of wastes for dumping in territorial seas. The MPRSA is also known as the Ocean Dumping Act ("Summary of the Marine Protection, Research, and Sanctuaries Act" 2018).

Perhaps the earliest law dealing very specifically with plastic materials was the Microbead-Free Waters Act of 2015. That act prohibited the manufacture, packaging, and distribution of so-called "rinse-off" consumer products containing plastic microbeads. Those materials were defined as any plastic material less than 5 mm in size. This action was applauded not only be environmentalists and others concerned with the problems of plastics pollution but also by representatives of the plastics industry itself, which pledged to find safe alternatives to micro-beads (see, for example, Embree 2015).

The most recent legislative effort to control ocean pollution by plastics was the Save Our Seas Act of 2018. That act amended and extended earlier legislation on ocean pollution and autho-rized an expenditure of $10 million per year, until 2022, to deal with this problem. Although enjoying wide bipartisan support, the act was criticized by some observers as being too tame an approach for such a serious problem. An effort to pro-duce a more comprehensive bill with greater consequences was proposed in 2019. Observers predicted that public perceptions of plastics issues had toughened since the adoption of the first version of the act, and they were uncertain about the chances of passage for the newer version of the bill (Toloken 2019).

The first piece of federal legislation to take a broad-based approach to virtually every aspect of plastics pollution was pro-posed in early 2020. That bill, the Break Free From Plastic Pol-lution Act, was first announced by Senator Tom Udall (D-NM) and Representative Alan Lowenthal (D-CA) in February 2020. The bill covered almost every imaginable problem associated with plastics, ranging from production to disposal. Among the major provisions of the bill are the following:

- Require producers of packaging, containers, and food-service products to design, manage, and finance waste and recycling programs
- Create a nationwide beverage container refund program
- Ban certain single-use plastic products that are not recyclable
- Ban single-use plastic carryout bags and place fee on the distribution of remaining carryout bags
- Establish minimum recycled content requirements for beverage containers, packaging, and food-service products
- Spur massive investments in U.S. domestic recycling and composting infrastructure
- Prohibit plastic waste from being shipped to developing countries
- Protect state and local governments that enact more stringent standards
- Place a temporary pause on new plastic facilities until EPA updates and creates important regulations on those facilities ("Pass the BFFP Pollution Act" 2020; for more details on the bill and a copy of the bill itself, see "Clark, Udall, Lowenthal, Merkley, Unveil Landmark Legislation to Break Free from Plastic Pollution" 2020).

Observers were not optimistic about the chances of passage of this bill. They pointed out that no Republican senators had signed on as co-sponsors of the bill, so its chances of success in a Republican-majority senate were poor.

In addition to these federal bills and acts, state and local governmental entities have also taken some actions on problems of plastic pollution. Most of these actions tend to deal with very specific, much more limited issues than does either SOS or BFFP Pollution Act. As noted earlier, eight states have some form of restriction on the use of plastic bags (although 14 states have adopted laws *preventing* local municipalities from adopting such laws) (Maldonado, Ritchie, and Kahn 2020). In addition, more than 300 cities and towns have adopted similar bans (Funkhouser 2019). Other states and towns have bans on other types of plastic materials and products. For example, Vermont has banned the use of plastic straws entirely, while California and Hawaii have placed restrictions on the use of such products. In addition, some major cities, several major corporations, and at least 23 countries around the world have also banned plastic straws (Woodward 2019).

## Conclusion

The role of plastics in modern society has mushroomed over the past seven decades or more. It is, indeed, almost impossible to imagine a world in which plastic materials and objects cannot be found everywhere. Yet, the spread of plastic technology has brought with it a host of environmental, health, social, political, economic, and other problems. The search for solutions to these problems has become a major focus of scientific and social research. That problem is only exacerbated, however, by the rapidly growing interest of fossil fuel companies in further expanding the production and use of plastics in everyday life. How these two trajectories will play out in the future is a question for which no answer is obviously available.

## References

Alabi, Okunola A., et al. 2019. "Public and Environmental Health Effects of Plastic Wastes Disposal: A Review." *Journal of Toxicology and Risk Assessment*. 5(1): https://doi .org/10.23937/2572-4061.1510021.

Alexander, Gemma. 2019. "How Sanitary Landfills Work." Earth911. https://earth911.com/business-policy/how -landfills-work/.

Al-Salem, S. M., P. Lettieri, and J. Baeyens. 2009. "Recycling and Recovery Routes of Plastic Solid Waste (PSW): A Review." *Waste Management*. 29(10): 2625–2643. https:// reader.elsevier.com/reader/sd/pii/S0956053X09002190

Al-Salem, S. M., P. Lettieri, and J. Baeyens. 2010. "The Valorization of Plastic Solid Waste (PSW) by Primary to Quaternary Routes: From Re-use to Energy and Chemicals." *Progress in Energy and Combustion Science.* 36(1): 103–129.

Antelava, Ana, et al. 2019. "Plastic Solid Waste (PSW) in the Context of Life Cycle Assessment (LCA) and Sustainable Management." *Environmental Management.* 64: 230–244. https://www.ncbi.nlm.nih.gov/pmc/articles/PMC6687704 /pdf/267_2019_Article_1178.pdf.

Anthony, Richard. 2013. "The Zero Waste International Alliance." In Paul H. Connett, ed. *The Zero Waste Solution: Untrashing the Planet One Community at a Time.* White River Junction, VT: Chelsea Green Publishing, 278–282.

Antony, Anu. 2017. "What Are Some of the Latest Waste-to-energy Technologies Available?" Prescouter. https://www .prescouter.com/2017/10/waste-to-energy-technologies -available/.

"Articles on Bioplastics." 2020. The Conversation. https:// theconversation.com/us/topics/bioplastics-14753.

Azoulay, David, et al. 2019. "Plastics & Health: The Hidden Costs of a Plastic Planet." Center for International Environmental Law, et al. https://www.ciel.org/wp-content /uploads/2019/02/Plastic-and-Health-The-Hidden-Costs -of-a-Plastic-Planet-February-2019.pdf.

Barrett, Axel. 2019. "Global Bioplastics Market to Grow by 20%." Bioplastics News. https://bioplasticsnews.com/2019 /08/08/global-bioplastics-market-to-grow-by-20/.

Bergman, Jennifer. 2011. "Ocean Gyres." Windows to the Universe. https://www.windows2universe.org/earth/Water /ocean_gyres.html.

"Biodegradable Plastics." n.d. Biodegradable Products Institute. https://www.bpiworld.org/page-190424.

Bläsing, Melanie, and Wulf Amelung. 2018. "Plastics in Soil: Analytical Methods and Possible Sources." *Science of the*

*Total Environment.* 612: 422–435. https://doi.org/10.1016 /j.scitotenv.2017.08.086.

Boots, Bas, Connor William Russell, and Dannielle Senga Green. 2019. "Effects of Microplastics in Soil Ecosystems: Above and below Ground." *Environmental Science & Technology.* 53(19): 11496–11506. https://doi.org/10.1021 /acs.est.9b03304.

Boswell, Clay. 2019. "Petrochemical Growing Pains for US Ethylene." Chemical Week. https://chemweek.com/CW /Document/102241/Petrochemicals-Growing-pains-for-US -ethylene.

Buranyi, Stephen. 2019. "The Missing 99%: Why Can't We Find the Vast Majority of Ocean Plastic?" The Guardian. https://www.theguardian.com/us-news/2019/dec/31/ocean -plastic-we-cant-see.

Cardoso, José. 2018. "The Circular Economy: Historical Grounds." In Ana Delicado, Nuno Domingos, and Luís de Sousa, eds. *Changing Societies: Legacies and Challenges: The Diverse Worlds of Sustainability.* Lisbon: Instituto de Ciências Sociais, Universidade de Lisboa, 115–128. https://www.researchgate.net/publication/328689953_The _circular_economy_historical_grounds.

"Centenary History of Waste and Waste Managers in London and South East England." 2007. Chartered Institution of Wastes Management. https://web.archive .org/web/20130813042213/http://www.ciwm.co.uk /web/FILES/About_CIWM/100_yrs_London_and_SE _centre.pdf.

"China Won't Accept U.S. Plastic Waste. Now What?" 2019. Statista. https://www.statista.com/chart/17220/plastic -waste-united-states/.

Choy, C. Anela, et al. 2019. "The Vertical Distribution and Biological Transport of Marine Microplastics across the Epipelagic and Mesopelagic Water Column." *Scientific*

*Reports*. https://www.nature.com/articles/s41598-019
-44117-2.pdf.

"Clark, Udall, Lowenthal, Merkley, Unveil Landmark
Legislation to Break Free from Plastic Pollution." 2020.
Katherine Clark. https://katherineclark.house.gov/2020
/2/clark-udall-lowenthal-merkley-unveil-landmark
-legislation-to-break-free-from-plastic-pollution
#:~:text=and%20U.S.%20Representative%20Alan%20
Lowenthal,reduce%20wasteful%20packaging%2C%20
and%20reform.

Collman, Ashley. 2019. "A Sperm Whale That Died on a
Scottish Beach Had 220 Pounds of Trash in Its Stomach."
Insider. https://www.insider.com/dead-sperm-whale-had
-220-pounds-of-trash-in-stomach-2019-12.

Cornwall, Warren. 2020. "To Replace Controversial Plastic
Additive BPA, a Chemical Company Teams up with
Unlikely Allies." Science. https://www.sciencemag.org
/news/2020/01/replace-controversial-plastic-additive-bpa
-chemical-company-teams-unlikely-allies.

"Cracking and Related Refinery Processes." 2016. Essential
Chemistry. https://www.essentialchemicalindustry.org
/processes/cracking-isomerisation-and-reforming.html.

"Crude Oil Fractional Distillation." 2020. 123RF. https://
www.123rf.com/photo_32867241_stock-vector-labeled
-diagram-of-crude-oil-fractional-distillation-.html.

Daigneau, Elizabeth. 2017. "Two Environmental Buzzwords.
Same Meaning?" Governing. https://www.governing
.com/topics/transportation-infrastructure/gov-zero-waste
-circular-economy.html.

Daly, Natasha. 2019. "Why Do Ocean Animals Eat Plastic?"
National Geographic. https://www.nationalgeographic.com
/animals/2019/12/whales-eating-plastic-pollution/.

d'Ambrières, Woldemar. 2019. "Plastics Recycling
Worldwide: Current Overview and Desirable Changes."

*Field Actions Science Reports.* Special Issue 19: 12–21. https://journals.openedition.org/factsreports/5102 #tocto2n5.

Delva, Laurens, et al. 2019. "Mechanical Recycling of Polymers for Dummies." Capture: Plastics to Resource. https://www.researchgate.net/publication/333390524 _AN_INTRODUCTORY_REVIEW_MECHANICAL _RECYCLING_OF_POLYMERS_FOR_DUMMIES.

Dermansky, Julie. 2020. "The Plastics Giant and the Making of an Environmental Justice Warrior." Desmog. https:// www.desmogblog.com/2020/01/07/formosa-sunshine -plastics-sharon-lavigne-environmental-justice.

Duncan, Emily M., et al. 2019. "Diet-Related Selectivity of Macroplastic Ingestion in Green Turtles (*Chelonia mydas*) in the Eastern Mediterranean." *Scientific Reports.* 9: 11581. https://doi.org/10.1038/s41598-019-48086-4.

"Effects of Microplastics on Fish and Invertebrates." 2018. Sixth International Marine Debris Conference. http:// internationalmarinedebrisconference.org/index.php/effects -of-microplastics-on-fish-and-invertebrates/.

Embree, Karl. 2015. "U.S. House Passes Legislation to Ban Plastic Microbeads." Plastics Today. https://www .plasticstoday.com/sustainability/us-house-passes-legislation -ban-plastic-microbeads/96124241823816.

"Entanglement." n.d. Blastic. https://www.blastic.eu /knowledge-bank/impacts/entanglement/.

"Entanglement: Entanglement of Marine Species in Marine Debris with an Emphasis on Species in the United States." 2014. National Oceanic and Atmospheric Administration, National Ocean Service. https://marinedebris.noaa.gov /sites/default/files/mdp_entanglement.pdf.

Enyou, Christian Ebere, et al. 2019. "Airborne Microplastics: a Review Study on Method for Analysis, Occurrence, Movement and Risks." *Environmental Monitoring and*

*Assessment.* 191(11): 1–17. Preprint available online athttps://www.preprints.org/manuscript/201908.0316/v1.

"Ethene (Ethylene)." 2020. Essential Chemistry. https://www .essentialchemicalindustry.org/chemicals/ethene.html.

"Ethylene Production in the United States from 1990 to 2018." 2020. Statista. https://www.statista.com/statistics /974766/us-ethylene-production-volume/.

Farah, Troy. 2020. "Bioplastics Continue to Blossom—Are They *Really* Better for the Environment?" Ars Technica. https://arstechnica.com/science/2020/01/are-bioplastics-all -hype-or-the-future-of-textiles/.

"Find Bag Legislation in Your Area." 2019. Bag the Ban. American Recyclable Plastic Bag Alliance. https://www .bagtheban.com/in-your-state/.

"Fishing's Phantom Menace: How Ghost Fishing Gear Is Endangering Our Sea Life." 2014. World Society for the Protection of Animals. https://www.worldanimalprotection .us/sites/default/files/media/us_files/sea-change-tackling -ghost-fishing-gear-report_us.pdf.

Fok, Lincoln, and P. K. Cheung. 2015. "Hong Kong at the Pearl River Estuary: A Hotspot of Microplastic Pollution." *Marine Pollution Bulletin.* 99(1–2): 112–118. https://www .researchgate.net/publication/280613449_Hong_Kong _at_the_Pearl_River_Estuary_A_hotspot_of_microplastic _pollution.

"Fueling Plastics: Series Examines Deep Linkages between the Fossil Fuels and Plastics Industries, and the Products They Produce." 2015. Center for International Environmental Law. https://www.ciel.org/reports/fuelingplastics/.

Funkhouser, David. 2019. "Banning Plastic Bags, Town by Town: A Guide." State of the Planet. https://blogs.ei.columbia .edu/2019/02/20/banning-plastic-bags-town-guide/.

Gardiner, Beth. 2019. "The Plastics Pipeline: A Surge of New Production Is on the Way." Yale Environment 360. https://

e360.yale.edu/features/the-plastics-pipeline-a-surge-of-new
-production-is-on-the-way.

Geyer, Roland, Jenna R. Jambeck, and Kara Lavendar Law.
2017. "Production, Use, and Fate of All Plastics Ever
Made." *Science Advances.* 3: e1700782. https://advances
.sciencemag.org/content/3/7/e1700782.

Gilmore, Nicholas. 2019. "The Imprudent Promise
of Plastics." Saturday Evening Post. https://www
.saturdayeveningpost.com/2019/06/the-imprudent
-promise-of-plastics/.

Gover, Jamison M., et al. 2019. Prey-size Plastics Are
Invading Larval Fish Nurseries." *PNAS.* 116(48): 24143–
24149. https://doi.org/10.1073/pnas.1907496116.

"The Graduate Script—Dialogue Transcript." n.d. Drew's
Script-O-Rama. http://www.script-o-rama.com/movie
_scripts/g/graduate-script-transcript-mike-nichols.html.

"Great Pacific Garbage Patch." 2020. National Geographic.
https://www.nationalgeographic.org/encyclopedia/great
-pacific-garbage-patch/.

"The Great Pacific Garbage Patch." 2020. The Ocean
Cleanup. https://theoceancleanup.com/great-pacific
-garbage-patch.

Grossman, David. 2018. "Plastic Bag Found at the Bottom
of the Mariana Trench." Popular Mechanics. https://www
.popularmechanics.com/science/environment/a20697886
/plastic-litters-even-the-very-bottom-of-the-ocean/.l

Hahladakis, John N., et al. 2018. "An Overview of Chemical
Additives Present in Plastics: Migration, Release, Fate and
Environmental Impact during Their Use, Disposal and
Recycling." *Journal of Hazardous Materials.* 344: 179–
199. https://www.sciencedirect.com/science/article/pii
/S030438941730763X.

Hayes, Douglas. 2019. "Micro- and Nanoplastics in
Soil: Should We Be Concerned?" Performance and

Adaptability: Biodegradable Mulch. https://ag.tennessee
.edu/biodegradablemulch/Documents/Microplastics-soil
-Factsheet-formatted.pdf.

"History of Waste Management." 2017. Recycle Guide.
https://www.recycleguide.org/history-waste-management/.

Hopewell, Jefferson, Robert Dvorak, and Edward Kosior.
2009. "Plastics Recycling: Challenges and Opportunities."
*Philosophical Transactions of the Royal Society of London B
Biological Sciences.* 364(1526): 2115–2126. https://doi.org
/10.1098/rstb.2008.0311.

Horan, Tegan S., et al. 2018. "Replacement Bisphenols
Adversely Affect Mouse Gametogenesis with Consequences
for Subsequent Generations." *Current Biology.* 28(18):
2948–2954.e3. https://doi.org/10.1016/j.cub.2018.06.070.

Horton, Alice C., et al. 2017. "Microplastics in Freshwater
and Terrestrial Environments: Evaluating the Current
Understanding to Identify the Knowledge Gaps
and Future Research Priorities." *Science of The Total
Environment.* 586: 127–141. https://doi.org/10.1016/j
.scitotenv.2017.01.190.

Houlihan, Jane, Sonya Lunder, and Anila Jacob. 2011.
"Timeline: BPA from Invention to Phase-out."
Environmental Working Group. https://www.ewg.org
/research/timeline-bpa-invention-phase-out.

Jamieson, A. J., et al. 2019. "Microplastics and Synthetic
Particles Ingested by Deep-Sea Amphipods in Six of the
Deepest Marine Ecosystems on Earth." *Royal Society Open
Science.* 6: 180667. https://doi.org/10.1098/rsos.180667.

Jeavans, Christine. 2008. "Plastic Recycling Comes Full
Circle." BBC News. http://news.bbc.co.uk/2/hi/uk_news
/magazine/7470662.stm.

Kaye, Leon. 2013. "3 Great Open Loop Recycling Projects."
Triple Pundit. https://www.triplepundit.com/story/2013/3
-great-open-loop-recycling-projects/53196.

Klinck, B. A., and M. E. Stuart. 1999. "Human Health Risk in Relation to Landfill Leachate Quality." British Geological Survey. http://citeseerx.ist.psu.edu/viewdoc /download?doi=10.1.1.475.1969&rep=rep1&type=pdf.

Koelmans, Albert A., et al. 2019. "Microplastics in Freshwaters and Drinking Water: Critical Review and Assessment of Data Quality." *Water Research*. 155(15): 410–422. https://doi.org/10.1016/j.watres.2019.02.054.

Kosuth, Mary, Sherri A. Mason, and Elizabeth V. Wattenberg. 2018. "Anthropogenic Contamination of Tap Water, Beer, and Sea Salt." *PLOS ONE*. 13(4): e0194970. https://doi .org/10.1371/journal.pone.0194970.

Kumar, Pramod. 2018. "Role of Plastics on Human Health." *Indian Journal of Pediatrics*. 85(5): 384–389.

Law, Kara Lavendar. 2017. "Plastics in the Marine Environment." *Annual Review of Marine Science*. 9: 205–229. https://doi.org/10.1146/annurev-marine-010816-060409.

"Laws that Protect Our Oceans." 2018. Environmental Protection Agency. https://www.epa.gov/beach-tech/laws -protect-our-oceans.

Leblanc, Rick. 2019. "The Decomposition of Wastes in Landfills." The Balance Small Business. https://www .thebalancesmb.com/how-long-does-it-take-garbage-to -decompose-2878033.

Lebreton, L., et al. 2018. "Evidence that the Great Pacific Garbage Patch Is Rapidly Accumulating Plastic." *Scientific Reports*. 8. Article number: 4666. https://doi.org/10.1038 /s41598-018-22939-w.

Lerner, Sharon. 2019. "Waste Only: How the Plastics Industry Is Fighting to Keep Polluting the World." The Intercept. https://theintercept.com/2019/07/20/plastics -industry-plastic-recycling/.

Lusher, Amy. 2015. "Microplastics in the Marine Environment: Distribution, Interactions and Effects." In

Melanie Bergmann, Lars Gutow, and Michael Klages, eds. *Marine Anthropogenic Litter*. Cham, Switzerland: Springer Open, 245–307. https://doi.org/10.1007/978-3-319 -16510-3_10.

Machado, Anderson Abel de Souza, et al. 2018. "Impacts of Microplastics on the Soil Biophysical Environment." *Environmental Science & Technology*. 52 (17): 9656–9665. https://doi.org/10.1021/acs.est.8b02212.

Maldonado, Samantha, Bruce Ritchie, and Debra Kahn. 2020. "Plastic Bags Have Lobbyists. They're Winning." Politico. https://www.politico.com/news/2020/01/20 /plastic-bags-have-lobbyists-winning-100587.

Mao, Jiude, et al. 2020. "Bisphenol A and bisphenol S Disruptions of the Mouse Placenta and Potential Effects on the Placenta-brain Axis." *Proceedings of the National Academy of Sciences of the United States of America*. 117(9): 4642–4652. https://doi.org/10.1073/pnas.1919563117.

Mason, Sherri A., Victoria G Welch, and Joseph Neratko. 2018. "Synthetic Polymer Contamination in Bottled Water." *Frontiers in Chemistry*. 6: 407. https://doi.org/10 .3389/fchem.2018.00407.

McCormick, Erin, et al. 2019. "Where Does Your Plastic Go? Global Investigation Reveals America's Dirty Secret." The Guardian. https://www.theguardian.com/us-news/2019 /jun/17/recycled-plastic-america-global-crisis.

McEachern, Kinsley, et al. 2019. "Microplastics in Tampa Bay, Florida: Abundance and Variability in Estuarine Waters and Sediments." *Marine Pollution Bulletin*. 148: 97–106. https://www.usfsp.edu/home/files/2019/09 /Microplastics-in-Tampa-Bay.pdf.

Messenger, Stephen. 2015. "Turtle Cut Free from 6-Pack Rings Is Unstoppable 20 Years Later." The Dodo. https://www.thedodo.com/turtle-six-pack-unstoppable -1166240209.html.

"Methodology for Allocating Municipal Solid Waste to Biogenic and Non-Biogenic Energy." 2007. Energy Information Administration. https://www.eia.gov /totalenergy/data/monthly/pdf/historical/msw.pdf.

"Microplastics in Drinking Water." 2019. World Health Organization. https://www.who.int/water_sanitation _health/publications/microplastics-in-drinking-water/en/.

Milner, Diane. 2017. "5 Most Common Industrial Chemicals." Noah Technologies. https://info.noahtech.com /blog/5-most-common-industrial-chemicals.

Moore, Charles. 2014. *Plastic Ocean: How a Sea Captain's Chance Discovery Launched a Determined Quest to Save the Oceans*. New York: Avery.

"Most Important Problem." 2020. Gallup. https://news.gallup .com/poll/1675/most-important-problem.aspx.

Nace, Trevor. 2018. "Here's a List of Every City in the US to Ban Plastic Bags, Will Your City Be Next?" Forbes. https:// www.forbes.com/sites/trevornace/2018/09/20/heres-a-list -of-every-city-in-the-us-to-ban-plastic-bags-will-your-city -be-next/.

"National Overview: Facts and Figures on Materials, Wastes and Recycling." 2019. Environmental Protection Agency. https://www.epa.gov/facts-and-figures-about-materials -waste-and-recycling/national-overview-facts-and-figures -materials#R&Ctrends.

"The New Plastics Economy: Rethinking the Future of Plastics." 2016. World Economic Forum. http://www3 .weforum.org/docs/WEF_The_New_Plastics_Economy .pdf.

"Oil: Crude and Petroleum Products Explained." 2019. U.S. Energy Information Administration. https://www.eia.gov /energyexplained/oil-and-petroleum-products/use-of-oil .php.

"101 Easy Eco Friendly, Zero Waste Tips." 2017. Going Zero Waste. https://www.goingzerowaste.com/blog/101-easy-eco-friendly-zero-waste-tips.

Parker, Laura. 2018. "The Great Pacific Garbage Patch Isn't What You Think It Is." National Geographic. https://www.nationalgeographic.com/news/2018/03/great-pacific-garbage-patch-plastics-environment/.

Parker, Laura. 2020. "An Old-School Plan to Fight Plastic Pollution Gathers Steam." National Geographic. https://www.nationalgeographic.com/science/2020/02/old-school-plan-to-fight-plastic-pollution-gathers-steam/.

Parton, Kristian J., Tamara S. Galloway, and Brendan J. Godley. 2019. "Global Review of Shark and Ray Entanglement in Anthropogenic Marine Debris." *Endangered Species Research*. 39: 173–190. https://www.int-res.com/articles/esr2019/39/n039p173.pdf.

"Pass the BFFP Pollution Act." 2020. Surfrider Foundation. https://www.surfrider.org/campaigns/introduce-bold-federal-legislation-to-tackle-the-plastic-pollution-crisis.

Pathak, Swati, C. L. R. Sneha, and Blessy Baby Matthew. 2014. "Bioplastics: Its Timeline Based Scenario & Challenges." *Journal of Polymer and Biopolymer Physics Chemistry*. 2(4): 84–90. https://doi.org/10.12691/jpbpc-2-4-5.

Pereira, Sabrina. 2019. "Plastic Perceptions: Surveying Public Opinion of Plastic Pollution in Rhode Island." University of Rhode Island Master's Theses. https://digitalcommons.uri.edu/theses/1480/.

Petre, Alina. 2020. "What Is BPA and Why Is It Bad for You?" Healthline. https://www.healthline.com/nutrition/what-is-bpa.

Phipps, Lauren. 2019. "The 5 Things You Need to Know about Chemical Recycling." GreenBiz. https://www

.greenbiz.com/article/5-things-you-need-know-about
-chemical-recycling.

"Plastic Manufacturing: Past, Present, and Future." 2020.
Craftech Industries. https://www.craftechind.com/plastic
-manufacturing-past-present-and-future/.

"Plastics: Material-Specific Data." 2019. Environmental
Protection Agency. https://www.epa.gov/facts-and-figures
-about-materials-waste-and-recycling/plastics-material
-specific-data.

"Plastics Recycling: Overview and Challenges." 2020. IFP
Energies Nouvelles. https://www.ifpenergiesnouvelles
.com/innovation-and-industry/our-expertise/climate-and
-environment/plastics-recycling.

"Public Concern about Plastic and Packaging Waste Is not
Backed up by Willingness to Act." 2018. Ipsos. https://
www.ipsos.com/ipsos-mori/en-uk/public-concern-about
-plastic-and-packaging-waste-not-backed-willingness-act.

Public Health Standing Committee. 2015. "Human Health
Impacts of Microplastics and Nanoplastics." NJDEP-
Science Advisory Board. https://www.state.nj.us/dep/sab
/NJDEP-SAB-PHSC-final-2016.pdf.

Qualman, Darrin. 2017. "Global Plastics Production, 1917 to
2050." Darrin Qualman. https://www.darrinqualman.com
/global-plastics-production/.

"Questions & Answers on Bisphenol A (BPA) Use in Food
Contact Applications." 2020. U.S. Food and Drug
Administration. https://www.fda.gov/food/food-additives
-petitions/questions-answers-bisphenol-bpa-use-food
-contact-applications.

Ralston, Brian E., and Tim A. Osswald. 2008. "The History
of Tomorrow's Materials: Protein-Based Biopolymers."
*Plastics Engineering*. 64(2): 36–40. https://cfpub.epa.gov
/ncer_abstracts/index.cfm/fuseaction/display.files/fileID
/14473.

Reed, Casey E., and Suzanne E. Fenton. 2013. "Exposure to Diethylstilbestrol during Sensitive Life Stages: A Legacy of Heritable Health Effects." *Birth Defects Research. Part C, Embryo Today: Reviews*. 99(2): 134–146. https://doi.org/10 .1002/bdrc.21035.

Rillig, Matthias C., et al. 2019. "Microplastic Effects on Plants." *New Phytologist*. 223: 1066–1070. https://doi.org /10.1111/nph.15794.

Rochman, Chelsea M., et al. 2016. "The Ecological Impacts of Marine Debris: Unraveling the Demonstrated Evidence from What Is Perceived." *Ecology*. 97(2): 302–312. https:// pdfs.semanticscholar.org/a14a/8a3bbf4aa24a60202c0c993 bbd8d0548a3eb.pdf.

Roegner, Eric. 2019. "What Is Closed Loop Recycling?" Amcor. https://www.amcor.com/about/media-centre/blogs /what-is-closed-loop-recycling.

Root, Tik. 2019. "Inside the Long War to Protect Plastic." The Center for Public Integrity. https://publicintegrity.org /environment/pollution/pushing-plastic/inside-the-long -war-to-protect-plastic/.

Royte, Elizabeth. 2019. "Is Burning Plastic Waste a Good Idea?" National Geographic. https://www.national geographic.com/environment/2019/03/should-we-burn -plastic-waste/.

Ruiz-Grossman, Sarah, and Damon Dahlen. 2017. "Heartbreaking Photos Show What Your Trash Does to Animals." HuffPost. https://www.huffpost.com/entry /plastic-trash-animals-photos_n_58ee9ec1e4b0b9e 984891ddf.

Sarena-Dilkes, Leele, et al. 2019. "Public Attitudes towards Plastics." *Resources, Conservation and Recycling*. 147: 227–235.

Savoca, Matthew S., et al. 2016. "Marine Plastic Debris Emits a Keystone Infochemical for Olfactory Foraging Seabirds."

*Science Advances.* 2(11): e1600395. https://doi.org/10.1126 /sciadv.1600395.

Savoca, Matthew S., et al. 2017. "Odours from Marine Plastic Debris Induce Food Search Behaviours in a Forage Fish." *Proceedings of the Royal Society B: Biological Sciences.* 284(1860). https://doi.org/10.1098/rspb.2017.1000.

Scheurer, Michagel, and Moritz Bigalke. 2018. "Microplastics in Swiss Floodplain Soils." *Environmental Science & Technology.* 52(6): 3591–3598. https://doi.org/10.1021/acs .est.7b06003.

Schwabi, Philipp, et al. 2019. "Detection of Various Microplastics in Human Stool: A Prospective Case Series." *Annals of Internal Medicine.* 171(7): 453–457.

"A Scientific Perspective on Microplastics in Nature and Society." 2019. Science Advice for Policy by European Academies. https://doi.org/10.26356/microplastics.

Sipper, Bill. 2019. "It's Time for Glass Again—Can We End Beverage Industry's Use of Plastic?" Waste Advantage. https://wasteadvantagemag.com/its-time-for-glass-again -can-we-end-beverage-industrys-use-of-plastic/.

Storrow, Benjamin. 2020. "Meet America's New Superpolluters: Plastic Plants." E&E News. https://www .eenews.net/stories/1062133995.

"Summary of the Marine Protection, Research, and Sanctuaries Act." 2018. Environmental Protection Agency. https://www.epa.gov/laws-regulations/summary-marine -protection-research-and-sanctuaries-act.

Teuten, Emma L., et al. 2009. "Transport and Release of Chemicals from Plastics to the Environment and to Wildlife." *Philosophical Transactions of the Royal Society B.* 364: 2027–2045. https://doi.org/10.1098/rstb.2008.0284.

Thiel, Martin, et al. 2018. "Impacts of Marine Plastic Pollution from Continental Coasts to Subtropical Gyres- Fish, Seabirds, and Other Vertebrates in the SE Pacific."

*Frontiers in Marine Science*. 5. https://doi.org/10.3389
/fmars.2018.00238.

Thompson, Richard C., et al. 2009. "Our Plastic Age."
*Philosophical Transactions of the Royal Society B*. 364: 1973–
1976. https://royalsocietypublishing.org/doi/pdf/10.1098
/rstb.2009.0054.

Toloken, Steve. 2019. "Save Our Seas 2.0 Bill Moves in
Congress but Faces Rough Waters." Plastic News. https://
www.plasticsnews.com/news/save-our-seas-20-bill-moves
-congress-faces-rough-waters.

Trimborn, Christel. 2020. "Galalith—Jewelry Milk Stone."
Ganoksin. https://www.ganoksin.com/article/galalith
-jewelry-milk-stone/.

Tullo, Alexander H. 2019. "Why the Future of Oil Is in
Chemicals, Not Fuels." *C&EN*. 97(8): 26–29. https://cen
.acs.org/business/petrochemicals/future-oil-chemicals-fuels
/97/i8.

"U.S. Chemical Industry Investment Linked to Shale
Gas Reaches $200 Billion." 2020. American Chemical
Council. https://www.americanchemistry.com/Media
/PressReleasesTranscripts/ACC-news-releases/US-Chemical
-Industry-Investment-Linked-to-Shale-Gas-Reaches-200
-Billion.html.

Vogel, Sarah A. 2009. "The Politics of Plastics: The Making
and Unmaking of Bisphenol A 'Safety.'" *American Journal of
Public Health*. 99(Suppl. 3): S559–S566. https://doi.org/10
.2105/AJPH.2008.159228.

Walch, Daniel C. 2002. "The Evolution of Refuse
Incineration." *Environmental Science & Technology*. 36(15):
316A–322A. https://pubs.acs.org/doi/pdfplus/10.1021
/es022400n.

Ward, Elizabeth. 2011. "Landfills a History." Green Risks.
http://greenrisks.blogspot.com/2011/07/landfills-history
.html.

Watts, Jonathan. 2019. "World's Deepest Waters Becoming 'Ultimate Sink' for Plastic Waste." The Guardian. https://www.theguardian.com/environment/2019/feb/27/worlds-deepest-waters-ultimate-sink-plastic-waste.

"What Is a Circular Economy?" 2017. Ellen MacArthur Foundation. https://www.ellenmacarthurfoundation.org/circular-economy/concept.

"What Is Advanced Recycling?" 2020. American Chemistry Council. https://plastics.americanchemistry.com/what-is-chemical-recycling/.

"What We Do." 2020. Gaia. https://www.no-burn.org/what-we-do/.

Woodward, Aylin. 2019. "In Some Countries, People Face Jail Time for Using Plastic Bags. Here Are All the Places That Have Banned Plastic Bags and Straws So Far." Business Insider. https://www.businessinsider.com/plastic-bans-around-the-world-2019-4.

Wright, Stephanie L., and Frank J. Kelly. 2017. "Plastic and Human Health: A Micro Issue?" *Environmental Science & Technology.* 51(12): 6634–6647. https://doi.org/10.1021/acs.est.7b00423.

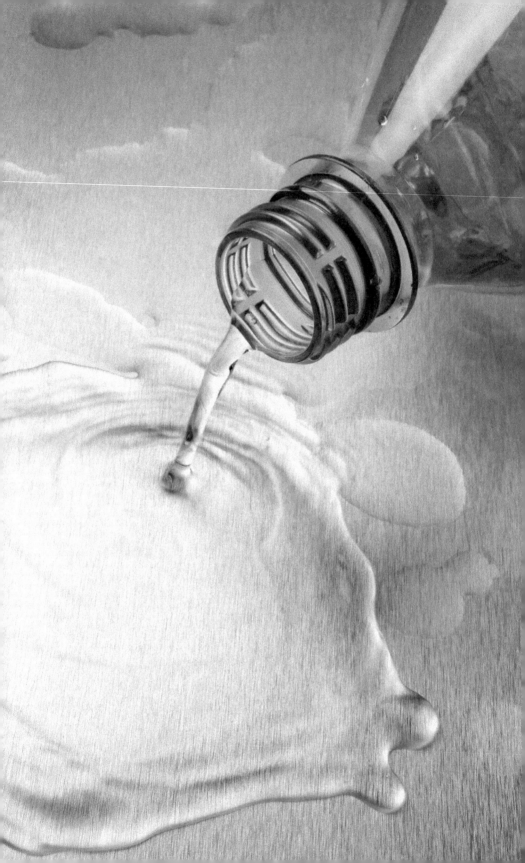

## Introduction

This chapter contains essays by individuals with special interest in the topic of plastics and microplastics. Some essays deal with current developments in the field, while others focus on a person's experiences with and thoughts about the problems of plastic pollution in today's world.

## Environmental Education That Empowers Students
*Ashlie Arkwright*

In 2018, 15-year-old Greta Thunberg from Stockholm, Sweden, captured the world's attention as she called for stronger action regarding climate change in front of her country's parliament. Like a rolling snowball gaining mass and momentum, her message garnered widespread support. In the same year, she passionately spoke at the United Nation's Climate Change Conference, and in the months that followed, millions of students joined her around the globe in protest. So far reaching was her influence that *Time* magazine named her their 2019 Person of the Year.

Greta Thunberg opted for a "school strike" to help spread her message about climate change. However, school itself can be

---

Bottled water is one of the world's fastest growing commercial beverages; however, its production depletes many environmental resources. By some estimates, as many as 1,500 plastic water bottles end up in landfills or the ocean every second of every day. (iStockPhoto.com)

an ideal setting for educating students about the environment, including the study of plastic pollution, and encouraging them to take a more active role in promoting change, such as reducing our consumption of single-use plastics. School-age students are naturally creative thinkers who are passionate about protecting our planet, and teachers play an important role in shaping their students' behavior and guiding them to become responsible citizens. In particular, teachers can use instructional time to accomplish three objectives in regard to environmental education. First, teachers can help students become environmentally literate by facilitating their acquisition of knowledge and understanding of and ability to effectively analyze and address complex environmental problems (Hollweg et al. 2011). Second, teachers can help their students become better stewards of our limited environmental resources, and third, they can empower students to become part of environmental solutions in their community. Given the number and magnitude of the global environmental issues we are facing, as well as the importance of raising up the next generation who will be charged with addressing those issues, environmental education that meets these three objectives should become a cornerstone of every school's curriculum.

But where to start? How do teachers transition from paying lip service to the importance of protecting our environment to nurturing the next generation of citizens who will develop solutions for the myriad environmental problems we currently face? Research studies that empirically isolate the features of programs responsible for measured outcomes are somewhat lacking; however, "active and experiential engagement in real-world environmental problems appears to be in favor with [environmental education] researchers and empirically supported" (Stern, Powell, and Hill 2014). Teachers can introduce students to complex environmental issues, such as our dependence on plastics and the myriad problems it creates due to lack of recycling and improper disposal, but it is also important to allow students the opportunity to grapple with these problems by working cooperatively to investigate them and propose solutions. Thus, after introductory teacher-directed instruction that lays a foundation about climate change and the role humans play in the process, educators should give teams of students the opportunity to choose specific issues that are personally meaningful to them. Teams can then research their issue of choice to understand it and its causes. Teachers can then charge teams with the task of creating and implementing innovative solutions that positively affect our environment. For example, a team of students could research the negative environmental impacts of improper disposal of single-use plastic bags. They might then decide to educate members of their community about these negative impacts through use of social media and by giving presentations to other groups of students. Additionally, they might collect single-use plastic bags and ensure that they are properly recycled, while also asking people to pledge to switch from single-use plastic bags to reusable plastic bags. They could even write letters to legislators encouraging them to create laws that limit the use of single-use plastic bags. Or they could organize a cleanup at a local stream. Student teams can then quantify the impacts of their efforts in terms of the number of followers on their social media sites, the number of peers to whom they present, the number of pledges they collect, the number of bags they recycle, or the number of pounds of garbage they remove from the environment. As an extension for older students, teachers can instruct and guide students to convert some of the data they collect into a reduction in carbon emissions.

Admittedly, this process requires a significant investment of instructional time; however, as a result, students will see the measurable fruits of their labor and who, like Greta Thunberg, will feel empowered to continue their efforts outside of the classroom and into the future. Margaret Mead famously said, "Never doubt that a small group of thoughtful, committed citizens can change the world; indeed, it's the only thing that ever has" (Brainy Quote 2020). Through intentional instruction, and by giving students the opportunity to take charge of their own learning, teachers have the opportunity to raise up the next generation of successful environmental leaders.

## References

Brainy Quote. 2020. "Margaret Mead Quotes." https://www .brainyquote.com/quotes/margaret_mead_100502.

Hollweg, K. S., et al. 2011. *Developing a Framework for Assessing Environmental Literacy*. Washington, DC: North American Association for Environmental Education. https://cdn.naaee.org/sites/default/files /devframewkassessenvlitonlineed.pdf.

Stern, Marc J., Robert B. Powell, and Dawn Hill. 2014. "Environmental Education Program Evaluation in the New Millennium: What Do We Measure and What Have We Learned?" *Environmental Education Research*. 20(5): 581–611. https://www.tandfonline.com/doi/full/10.1080 /13504622.2013.838749.

*Ashlie Arkwright, PhD, has been teaching sixth-, seventh-, and eighth-grade integrated science at the School for the Creative and Performing Arts at Bluegrass in Lexington, Kentucky, for the past 20 years.*

## Microplastics and Plankton
*Cherilyn Chin*

Plastics in the marine environment are a critical problem. Large pieces of plastic can entangle and otherwise harm ocean wildlife, and smaller pieces can get ingested and cause a whole host of problems, including impaction and starvation. Even more insidious are the microplastics that are less than 5 mm in diameter. Microplastics cause serious problems for plankton, which includes phytoplankton, the microscopic plants of the ocean, and zooplankton, the microscopic animals of the sea. Both form the base of the food chain in the ocean.

As plants, phytoplankton do not ingest microplastics in the same active way as zooplankton. Instead, microplastics stick to phytoplankton, and this affects organisms high on the food chain that eat it. In all marine organisms, microplastics can bioaccumulate. Bioaccumulation is building up of a substance in an organism over time. Larger creatures, such as fish or filter-feeding whales and sharks, eat the smaller creatures, such as plankton. Over time, each step up on the food chain causes animals to accumulate more microplastics over time in their bodies (Bergmann, Gutow, and Klages 2015, 125). Microplastics are so abundant in the sea that humans are now eating them in their seafood (Smith et al. 2018).

Microplastics sticking to phytoplankton also matter because they are a part of the marine snow. Marine snow consists of organic particles from the surface that sink into deeper waters and the sea floor. These organic particles may include dead plankton, feces, and matter from the land. They bring food, in the form of organic matter, from the surface down into the water column and onto the ocean floor for deep-sea animals to eat. Depending on their size, microplastics have an impact on how fast the phytoplankton sink (Long et al. 2015, 40): too fast and zooplankton do not have time to eat the phytoplankton; too slow and the phytoplankton do not make it to the ocean floor as marine snow.

Microplastics cause numerous problems for zooplankton. Microplastics can stick to zooplankton as well as to phytoplankton. In a study, copepods, a kind of fast-moving zooplankton, were found to have microplastics stuck to their feeding appendages and legs (Cole et al. 2013, 6653). So animals that eat copepods not only ingest whatever microplastics the copepods have eaten and internalized, but also whatever is externally stuck to them. The animals then bioaccumulate more microplastics.

The ingestion of microplastics can cause many problems in zooplankton. Ingested microplastics can impact the physiology of zooplankton, including interfering with the acquisition of food and impacting the gut as well as causing starvation (Land 2015, 4). In a study of copepods, microplastics were found to interfere with their feeding, so they ate less. Scientists found that the copepods avoided eating algae that was similar in shape and size to the microplastics that they were exposed to (Coppock et al. 2019, 785). This could have implications on their health if they are actively avoiding microplastic-sized food. Ingested microplastics also had an impact on reproduction. The copepods had smaller eggs, put less time into reproduction, and had decreased hatching success (Cole et al. 2015, 1135).

The fecal pellets that copepods excrete are full of microplastics. Fecal pellets with microplastics act differently in the water column than regular fecal pellets that are not full of microplastics. The fecal pellets with microplastics break up faster and sink more slowly (Coppock et al. 2019, 787). The slow sinking increases the opportunity for the fecal pellets to be eaten by organisms in the water column, but it decreases the amount of organic matter reaching the ocean floor. The long-term implications of these changes are unknown.

Krill have been found to break down consumed microplastics into smaller nanoplastics (less than 1 μm) (Dawson et al. 2018). Krill can bioaccumulate nanoplastics or excrete them. The effect of these smaller particles may be the same or different than microplastics on organisms. They can transport these excreted nanoplastics from one area to another depending on the krill's fate in the food chain. A krill that is eaten can bring micro- or nanoplastics to new areas of the ocean.

Why should you care about tiny pieces of plastic in the ocean? Microplastics are everywhere in the ocean, whether at the surface, in the deep sea, or on the ocean floor. They are toxic to most organisms, whether it is chemicals leaching out of them or chemicals sticking to them or when ingested (even by humans). Scientists are investigating in both the laboratory and the field to figure out the specific effects of microplastics' fate in the ocean and its implications for the marine food chain in the future.

## References

Bergmann, M., L. Gutow, and M. Klages, eds. 2015. *Marine Anthropogenic Litter*. Berlin, Germany: Springer.

Cole, M., et al. 2013. "Microplastic Ingestion by Zooplankton." *Environmental Science & Technology*. 47(12): 6646–6655.

Cole, M., et al. 2015. "Impact of Polystyrene Microplastics on Feeding, Function and Fecundity in the Marine Copepod Calanus helgolandicus." *Environmental Science & Technology*. 49(2): 1130–1137.

Coppock, R. L., et al. 2019. "Microplastics Alter Feeding Selectivity and Faecal Density in the Copepod, Calanus helgolandicus." *Science of the Total Environment*. 87: 780–789.

Dawson, A. L., et al. 2018. "Turning Microplastics into Nanoplastics through Digestive Fragmentation by Antarctic Krill." *Nature Communications*. 9(1). https://doi.org/10.1038/s41467-018-03465-9.

Land, M. 2015. "Effects of Nano- and Microplastic Particles on Plankton and Marine Ecosystem Functioning." EviEM. https://www.researchgate.net/publication/297214292_Effects_of_nano-_and_microplastic_particles_on_plankton_and_marine_ecosystem_functioning.

Long, M., et al. 2015. "Interactions between Microplastics and Phytoplankton Aggregates: Impact on Their Respective Fates." *Marine Chemistry*. 175: 39–46.

Smith, M., et al. 2018. "Microplastics in Seafood and the Implications for Human Health." *Current Environmental Health Reports*. 5: 375–386.

*Cherilyn Chin is a marine biologist, ocean conservationist, writer, blogger at* Ocean of Hope, *and photographer. Her life's purpose is to bring to light the plight of our oceans and to reconnect people to nature.*

## Microplastics Are Everywhere: Threatening the Health of the Planet and Future of Humanity
*Mark Friedman*

According to the United Nations, 51 trillion microplastic particles are in our environs. Research published in *Scientific American, Science, Marine Pollution Bulletin* (Andrady 2011; Jambeck 2015; Thompson 2018) and other national and international scientific journals shows just how prevalent they are. They've been found in Arctic sea ice, bottled water (the bottle being one of the chief sources of single-use plastic), beach sand, and high atmosphere air samples. Plastic fibers and particles are found on the tallest mountains and deepest ocean floors. They are ubiquitous in our food chain. Absent action by all of us, continued life on earth is problematic.

Every year, about 8.8 million tons of plastic are dumped into the ocean ("The Facts" 2020). Already, it is devastating marine life (from whales to plankton) and, ultimately, will cause marine ecosystem collapse. Scientists have found that the sun's ultraviolet rays break the plastic down into chemicals that mimic estrogen, disrupting human hormonal systems, especially in adolescents and pregnant women ("Hormone Disrupting Chemicals & Consumer Products" 2016).

Also, bisphenol A (BPA), used to make hard plastic, is released by plastic breakdown. Items containing BPA include plastic bottles, medical equipment, toys, consumer electronics, household appliances, and automobiles. Epoxy resins containing BPA are used in liners of food and beverage cans and thermal cash register receipts. BPA, already known to be hazardous, is banned from baby bottles ("Hormone Disrupting Chemicals & Consumer Products" 2016). Surprisingly, plastic is also released from clothing, especially fleece, as it is made from synthetic materials. When washed, the clothes shed plastic fibers: an estimated minimum of 1,900 plastic microfibers each washing (Thompson 2018).

In the ocean, plastic acts like a sponge, picking up toxins and chemicals. Every marine organism that has been studied contains plastic—even microscopic plankton. Microplastics are then ingested by marine organisms that we commonly eat, such as shellfish and bony fish. Studies have found up to 84 pieces of plastic in individual bony fish (Andrady 2011).

Students all over the world have become aware of the dangers of microplastics. Japanese marine biology teacher Yasuyuki Kosaka initiated research collaboration on ocean microplastics in 2017 with the Los Angeles Ánimo High School marine biology club and this author and club mentor. This collaboration broadened to include students from more than half a dozen countries and teachers from a dozen others.

Using standard scientific method, Japanese students were able to flush microplastic particles from the stomachs of oysters in the Sea of Japan—a seafood supply source (aquaculture and fishing) for Kyoto.

In Los Angeles, we created the Los Angeles Microplastics Team. The students teamed up with the Los Angeles Maritime Institute to conduct research from tall sailing ships using a manta trawl donated by 5Gyres. They also collected water and sand samples from Los Angeles harbors and beaches, finding tens of thousands of plastic nodules and macro- and microplastic debris and filaments.

The team leads tall ship passengers on monthly bilingual hands-on, onboard research and data collection expeditions thanks to a grant from the California Coastal Commission. The focus is environmental education of the Spanish-speaking community.

The team made numerous presentations at other high schools and colleges, which led to the formation of new environmental and marine biology clubs to carry out their own research and post the data and research on a student-developed international, multilingual website. Students also presented their research at local, regional, and international science fairs, while organizing to have their action campaign for solving the microplastics problem spread widely. These activities have created a passion in the team members to educate others to make changes in their lives that will reduce the growth of microplastic pollution, and they have also taught them communication and cooperation skills necessary to the process.

Solutions the students have put forward are as follows:

- Reduce food packaging.
- Recycle existing plastic.
- Make new plastic products recyclable.
- Adopt paper, bamboo, cornstarch, and other carbohydrates as biodegradable substitutes for plastic.
- Make corporations pay for the cleanup and recycling costs of their plastics.
- Force governments to implement plastic reduction, recycling campaigns, and job-creating innovations.
- Advocate for more stringent and enforced environmental regulations.

Los Angeles students along with Latinx youth from around the United States attended and presented at Cubambiente, the international environmental conference in Havana, Cuba, collaborating with the Jovenes Ambientales Cubanos, Acuario Nacional de Cuba, and Universidad de Habana. The team expanded their international collaboration through an Asia-wide youth environmental conference in Japan.

The students have met, via Skype, with youth in Japan, Taiwan, Honduras, England, India, Chile, Cuba, and Indonesia to initiate local research on microplastics, expanding education and action campaigns. The team also collaborates with non-profit organizations and aquaria. Their material and resources are available free at www.lamitopsail.org.

The Microplastics Team captain Johanna Cervantes said she took with her valuable experiences from the collaboration. "I have learned to be a better communicator and team player,"

Cervantes said. "Thanks to this overseas collaboration I've also learned how to communicate with different people and how to properly get my messages across."

Teammate Jose Velasquez had a similar experience. "We realized that there is little awareness in our community on plastic pollution," he said. "We collected samples of microplastics at nearby beaches to prove that microplastic pollution is a problem that directly affects marine organisms and humans." Kevin Zepeda added, "That initial passion to create awareness allowed us to present our research at science fairs to educate individuals from our community to make changes in their lives that will reduce the growth of microplastic pollution."

Worldwide, most people understand that the crises we face with plastic pollution, species extinction, ocean acidification, and impacts of climate change are inextricably linked. Innovations and solutions already exist. What is lacking is the political will by governments and corporations to carry out necessary measures to begin to rectify these problems. Perhaps we can follow Cuba's lead. Cuba has a 100-year plan to respond to the impacts of climate change and other environmental crises like ocean pollution on their food supply, tourist industry, jobs, and health care.

It is up to the new generation of youth, such as Swedish climate activist Greta Thunberg, to organize student strikes and street protests and challenge the profit system. Some have attacked her for her critique of capitalism and corporate greed and for advocating activism and leading student strikes. Her eloquent response is, "We are fighting for everyone's future and, if you think we should be in school instead, then we suggest that you take our place in the streets striking from your work. Or better yet, join us, so it can speed up the process" (Speech to European Economic and Social Committee, Brussels, February 21, 2019).

### References

Andrady, Anthony L. 2011. "Microplastics in the Marine Environment." *Marine Pollution Bulletin.* 62(8): 19–22.

"The Facts." 2020. Plastic Oceans. https://plasticoceans.org
/the-facts/.

"Hormone Disrupting Chemicals & Consumer Products."
2016. Physicians for Social Responsibility. www.psr.org.

Jambeck, Jenna R. 2015. "Plastic Waste Inputs from Land
into the Ocean." *Science*. 347(6223): 768–771.

Thompson, Andrea. 2018. "From Fish to Humans, a
Microplastic Invasion May Be Taking a Toll." Scientific
American. https://www.scientificamerican.com/article/from
-fish-to-humans-a-microplastic-invasion-may-be-taking-a
-toll/.

*Mark Friedman is mentor to the Los Angeles Microplastics Team.
Teachers and student environmental clubs interested in collaborating on microplastic research, educational, and action campaigns
should contact him at Marklewisfriedman@gmail.com.*

## How to Be More Eco-conscious
*Pratibha Gopalakrishna*

"Would you like a bag for ten cents?" the cashier asks as I realize I forgot my tote bag. The bulk of withheld guilt pours in. A
similar situation occurs every time I buy a beverage, especially
one requiring a straw. The guilt-tripping ends with the same
question: When will I be more eco-conscious?

Environmental concern grew during the late twentieth century, when scientists observed how various pollutants affected
earth's soil, rivers, and oceans. Increased chemical pesticide use,
carbon-fuel consumption, oil spills, and dumping wastes into
waterways were identified as the initial culprits that caused
leaching. One among these pollutants was plastic, which, to
this day, still remains an urgent threat.

Plastic pollution is a nebulous presence we sometimes choose
to ignore. Plastic is present everywhere: toothbrush, nonstick

pans, water bottles, cosmetics, automobile parts, stationeries—the list is endless. Picturing a world without plastics is not just hard; it's impossible.

Plastics were invented out of need and because natural resources started depleting. Around the 1860s, the elephants of the world were in danger of becoming extinct because they were hunted for their ivory tusks, which were used to make billiard balls. John Wesley Hyatt, a young artisan, invented a crude version of plastic from cellulose, calling it "celluloid." Though celluloid didn't end up on the billiards table due to its brittle nature, it infiltrated other common household items such as combs, toothbrushes, and photographic films.

The first truly synthetic plastic called Bakelite was invented by Leo Baekeland, a Belgian chemist, in 1907. This invention sparked interest among scientists all around the world to invent new materials rather than copy natural elements. Broadly defined as polymers, plastics constitute a molecular chain of carbon, hydrogen, oxygen, and silicon. These molecules come from breaking down crude oil, petroleum, and other products. Varying the combination changes the characteristics of the plastic, making it strong and durable or weak and fragile.

Plastic gave way to a lot of other inventions that made life easier. Its malleable nature was an instant hit. The early twentieth century saw a deluge of plastic materials infiltrating every household in the United States, and plastic production increased after World War II. Plastic also helped protect much of the natural resources that were initially exploited, such as wood, ivory, paper, metals, and mollusk shells. From toys to furniture, plastic became ubiquitous until it ended up becoming the worst form of pollution humanity has faced.

Around the 1970s, scientists noticed that plastics did not degrade as natural polymers and remained in the environment. Plastic debris in soils and oceans and health hazards due to bisphenol A (BPA) in plastics were recorded. Evidence started mounting against plastic's undegradable nature, and researchers asked the world to pay heed.

Forty years later, plastic pollution still haunts us, and researchers have found that these plastics break down to smaller parts called microplastics, whose ill effects are yet unknown. As the name implies, microplastics are tiny plastic pieces ranging from a few millimeters to micrometers, such as the tiny beads found in toothpaste and facewash. These microplastics enter the ocean where fish and mollusks consume them, or they form sediments in lakes and ponds, interfering with worm habitats. Although their effects on humans are still being researched, microplastics are a growing concern.

With all this in mind, how do we be more ecologically conscious?

My answer is individual action. Our choice to consciously use plastic can either kill a turtle (or any other land and ocean dweller) or save one. Nothing can be done about the plastics already present in the environment, unless we invent a way to find, collect, and properly recycle them. The best option we have right now is to reduce and reuse.

Firstly, there are numerous ways to reduce plastic consumption. Bringing your own spoon-fork-straw set, mugs when heading out for coffee, or lunch boxes for packing up leftovers after a hearty meal—these actions can compound and people appreciate the gesture. Carrying a bag while grocery shopping saves time and money, and it feels good. Plus, these actions can prompt interesting conversations and inspire others. Nobody can make fun if you reply, "I'm doing my part for the environment." Tiny changes make a huge difference.

Secondly, I cannot stress more on reusing plastics. Currently, sold plastics are more durable, washable, and reusable—at least twice or thrice. They can be used as garbage bags and storage bags or used to hold lunch boxes in backpacks. We make use of these plastics as much as possible before dumping them in the recycling bin.

On the contrary, dumping plastics into the recycling bin does not count as an individual action because a recent report by Greenpeace, a nonprofit organization, shows that most of the plastics labeled as "recyclable" are not actually recycled. Instead, they end in landfills or incineration facilities. Moreover, recycling plants in the United States can't process most of the plastic wastes generated, which means these wastes also end up in landfills, eventually finding their way to the oceans.

Finally, create awareness about plastic waste by practicing the two Rs: reduce and reuse. Individual actions have twofold results: they form habits and provide inspiration to others. Consistent practice forms habits, and while communication is effective, it can't beat action that hits the sensory system and leaves an imprint on the brain.

So, I reach home and place a grocery bag in all my backpacks, along with a small, portable bag containing a spoon-fork-straw set and a mug. All that's left to do is carry them around without fail.

*An avid storyteller with a passion for science, music, and art, Pratibha Gopalakrishna went from studying cellular processes through microscopes to writing about them. You can read her words at protonjump.com.*

## Biological Recycling of Polyethylene Shopping Bags
*Joel Grossman*

Greater wax moth, *Galleria mellonella*, caterpillars are the best species known when it comes to eating, digesting, and, thereby, biologically recycling polyethylene plastic shopping bags. These bags are currently banned by many jurisdictions for supposedly being nonbiodegradable and lasting up to a thousand years in the environment. Notorious for wreaking havoc in honey bee hives, wax moth caterpillars are tiny, armored, tunneling machines with retractable antennae and three powerful front teeth for ripping through beeswax and honeycomb. A closely related moth pest commonly infesting kitchen cabinets and stored products worldwide, Indian meal moth, *Plodia interpunctella*, also eats polyethylene, as do myriad fungi, bacteria, aquatic algae, and other organisms.

It is perhaps ironic that "pests" routinely fumigated with highly toxic gases might one day be "repurposed" in "biofactories" to ameliorate humanity's annual 500 million to 1 trillion polyethylene plastic shopping bag consumption binge. "Usually, when the word 'polyethylene' is mentioned in the context of discussing sustainability issues, a good chance the message is that 'the current level of environmental plastic pollution is unsustainable,'" writes Svetlana Boriskina. "Polyethylene does indeed comprise a large volume of plastic waste" and "is one of the most-produced materials in the world" for "many good reasons," including its "unique material properties" (Boriskina 2019).

"Some of the emerging applications of polyethylene hold high promise for sustainable energy generation from renewable sources and for sustainable management of planetary energy and water resources," said Boriskina. "Polyethylene offers an excellent alternative to metals, providing corrosion resistance, higher strength per unit weight, and high thermal conductivity combined with electrical insulation." Besides being an alternative to mining the earth for metals, polyethylene is "an excellent alternative to glasses, combining light weight, flexibility, and optical transparency not only in the visible but also in the infrared spectral range." In the textile industry, polyethylene fibers are considered sustainable, provided a recycling solution is in place.

Polyethylene is many (poly) ethylene ($CH_2=CH_2$) molecules hooked together, in essence long linear strings of carbon atoms studded with hydrogen atoms—carbon and hydrogen, nothing more. "No element is more essential to life than carbon, because only carbon forms strong single bonds to itself that are strong enough to resist chemical attack under ambient conditions," writes chemist John Emsley. "This gives carbon the ability to form long chains and rings of atoms, and these are the structural basis for many compounds that comprise the living cell" (Emsley 2003).

"What allows the wax worm to degrade a chemical bond not generally susceptible to bio-degradation?" ask researchers Paolo Bombelli and his colleagues. The answer may lie in the ecology of the wax worm itself. "They feed on beeswax. . . . The most frequent hydrocarbon bond is the C-C, as in polyethylene. . . . It seems likely that the C-C single bond . . . is one of the targets of digestion" (Bombelli et al. 2017). In a shopping bag experiment with wax moth caterpillars, there was "a significant mass loss of 13 percent polyethylene over 14 hours of treatment . . . which is markedly higher than the rate of PET biodegradation by a microbial consortium" (Bombelli et al. 2017). This mechanism also works with polyethylene terephthalate (PET), a plastic packaging material. Crudely removing caterpillar digestive enzymes, some of which may be produced by microbes inside our gut, and smearing the mush onto polyethylene bags also produces biodegradation.

One by-product of polyethylene digestion by wax moth caterpillars is ethylene glycol, better known as antifreeze, a highly toxic material. Over time, wax moth caterpillar survival declines on a polyethylene diet as ethylene glycol increases. At the 2019 Entomological Society of America annual meeting in St. Louis, Harald Grove and colleagues from Manitoba, Canada's Brandon University, cited this as additional evidence that polyethylene shopping bags were being digested by wax moth caterpillars, not just shredded by their powerful jaws (Grove et al. 2019).

Unshackled from the "false belief" that polyethylene is inherently nonbiodegradable, a world of potential biological recycling solutions opened up. Hundreds of common bacteria species producing enzymes biodegrading (digesting) plastics were cataloged by Sen and Raut (2015). In incubation experiments, a heat-loving soil bacteria, *Brevibacillus borstelensis*, reduced polyethylene weight by as much as 30 percent.

When Pramila and Ramesh (2011) searched landfills in India, two common fungi, *Mucor circinilloides* and *Aspergillus flavus*, were hard at work destroying polyethylene shopping bags. Underwater in India, "microalgae like green algae, blue-green algae and diatoms were isolated from the domestic polyethylene bags dumped in the suburban water bodies and studied for its potency on deterioration of polyethylene," said Kumar et al. (2017). "The most dominant microalgae were *Scenedesmus dimorphus* (green microalga), *Anabaena spiroides* (blue-green alga) and *Navicula pupula* (diatom). . . . A better solution for the complete degradation of polyethylene has not yet been formulated."

"Rapid biodegradation is the only eco-friendly process which can solve the problem," said researchers Nupur Ojha and colleagues. *Penicillium fungi*, known as fruit-infesting molds, as starter cultures to ripen blue-mold cheeses and for production of the antibiotic penicillin, also possess enzymes which "have plastic degrading abilities." One polyethylene recycling scenario involves chaining together multiple microbe species: "The breakdown of large polymers to carbon dioxide (mineralization) requires several different organisms, with one breaking down the polymer into its constituent monomers, another one being able to use the monomers and excreting simpler waste compounds as by-products and a final one being able to use the excreted wastes." These polyethylene-destroying fungi "can be further enhanced in an industrial scale for degrading various plastic materials" (Ojha et al. 2017).

Mass production of fungi, bacteria, algae, earthworms, and insects for biological pest control, waste composting, and food and pharmaceutical production is almost routine. Wax moth caterpillars are grown as a "model organism" and ethical alternative for medical researchers studying the immune system and developing therapies against human diseases. Indeed, wax moth caterpillars are a common item of commerce, and setting up a science experiment to test the biodegradability of polyethylene shopping bags is just a few Internet key taps away at scientific supply houses, bait shops, Petco, Grubco, eBay, Amazon, Walmart, and so on. But don't believe the "90-day returns" banner.

## References

Bombelli, Paolo, et al. 2017. "Polyethylene Bio-degradation by Caterpillars of the Wax Moth *Galleria mellonella.*" *Current Biology.* 27(8): R292–R293. https://www.sciencedirect.com/science/article/pii /S0960982217302312.

Boriskina, Svetlana V. 2019. "An Ode to Polyethylene." *MRS Energy & Sustainability.* https://www.cambridge .org/core/journals/mrs-energy-and-sustainability/article /an-ode-to-polyethylene/9A102B14FFA53554B629695EC E0A0676.

Emsley, John. 2003. *Nature's Building Blocks: An AZ Guide to the Elements.* New York: Oxford University Press.

Grove, Harald C., et al. 2019. "Plastic Degradation in Caterpillars of the Greater Waxworm (*Galleria mellonella*) and the Route to Metabolic Effects." Entomology. https:// esa.confex.com/esa/2019/meetingapp.cgi/Paper/148518.

Kumar, R. V., et al. 2017. "Biodegradation of Polyethylene by Green Photosynthetic Microalgae." *Journal of Bioremediation and Biodegradation.* 8(381): 2. https:// www.researchgate.net/profile/Rajesh_Gopal5/publication /312590130_Biodegradation_of_Polyethylene_by_Green _Photosynthetic_Microalgae/links/5c2b94fa458515a 4c7058654/Biodegradation-of-Polyethylene-by-Green -Photosynthetic-Microalgae.pdf.

Ojha, Nupur, et al. 2017. "Evaluation of HDPE and Ldpe Degradation by Fungus, Implemented by Statistical Optimization." *Scientific Reports.* 7: 39515. https://www .nature.com/articles/srep39515.

Pramila, R., and K. Vijaya Ramesh. 2011. "Biodegradation of Low Density Polyethylene (LDPE) by Fungi Isolated from Municipal Landfill Area." *Journal of Microbiology and Biotechnology Research* 1(4): 131–136. https://pdfs .semanticscholar.org/7451/c723b8ce9c3bfd0c9e0eafcb69f9 376c7365.pdf.

Sen, Sudip Kumar, and Sangeeta Raut. 2015. "Microbial Degradation of Low Density Polyethylene (LDPE): A Review." *Journal of Environmental Chemical Engineering.* 3(1): 462–473. https://www.sciencedirect.com/science /article/pii/S2213343715000056.

*Joel Grossman has worked as an entomologist. He writes the Bio-control Beat blog.*

## Impacts of Plastic on Sea Turtles
*Jake Krauss*

Walking along the beach, the desert sun burns hot, yet the waves bring water much too cold for my feet. The fresh ocean smell of the Gulf of California is momentarily interrupted by a rotting, dead sea turtle on the beach. Like many others, this individual most likely was caught in a loose fishing net and drowned. Alternatively, it may have eaten one too many plastic bags floating in the bay, confusing the shimmering plastic with jellyfish, a common food item for the turtles.

I'm brought back to my first encounter with a sea turtle here in Kino Bay, Mexico. A small beak is visible scratching the water's surface, a sea turtle taking a gulp of breath while caught in a net. The creature is taken out and released into the boat, and it flaps its smooth, leathery fins frantically about. My big toe is crushed under the weight of her shell as she scuttles across the boat. The moment passes, and the turtle calms down, lying perfectly still in submission. Limpets, parasites that grow specifically on sea turtles, line its face, and the boat captain cuts them off carefully with a knife. The captain is part of Grupo Tortuguero, a turtle-monitoring group that assesses the population in this part of the Gulf to help in their conservation. Overall, this group has caught and released 600 turtles, an effort that has been recognized as contributing significantly toward conservation.

The sea turtle is brought back to the beach, where youth from the Indigenous Comcaac community wait patiently. Out on the beach, the turtle is measured, weighed, and given a health assessment while the boat captain explains the process. Lifting up one end of the stick used to weigh the turtle, I struggle to keep the 46 kg animal aloft while measurements are taken. The firm shell and beak that made the turtle seem so fierce suddenly give way to a slight vulnerability as the soft tail and flippers are exposed. How long will it take, I wonder, before this large beast consumes enough plastic bags and she ends up back on the beach to rot?

Outreach efforts such as this help communities to develop an appreciation for the local wildlife and understand the challenges they might face. Indeed, part of finding solutions to the problems of plastic pollution hinges on community support. For example, the Prescott College Kino Bay Center organizes an annual cleanup of the site where we caught the sea turtles, involving many different community groups. In addition to removing the trash, the activity serves as an awareness campaign for the effects of trash on the local wetland, Laguna la Cruz, and can inform future changes in behavior. This is one of the many outreach activities conducted with the community to encourage sustainability and environmental stewardship.

### In Far-Flung Places

The reach of plastics extends far beyond the beaches and places where people roam. I have the chance to go to one of the Midriff islands in the Gulf of Mexico, rocky volcanic formations erupting out of the cerulean waters. Scaling the crumbling mountainside, I summit to catch a glimpse of double-breasted cormorants nesting on the rocky precipice. Carefully hauling myself up limb by limb, I finally make it to the ridge top to perch and watch the scene below. Hundreds of nests and seabirds cover the valley, and I spy a pair of cormorant fledglings waking up to feed. I am dismayed when I notice a piece of plastic wound in this nest, in a place thought to be unreachable by humans. It is at this moment I truly understand the importance of the outreach work done to reduce plastic pollution.

## Entanglement

The local community of Kino is a fishing-based economy, dependent on the health of the sea. However, many of the fishing habits are unsustainable, with pollution from fishing equipment being a major contributor to entanglement with marine wildlife. This can also affect animals as large as whales; they can consume the loose fishing lines, which lodge in the digestive system and cause blockages. Eventually, the animal could starve to death. In my time here, I've come across a huge blue whale washed up on the beach. As to the cause of death, it's hard to say, with the multicolor flesh-feeding flocks of vultures and pelicans, but I can't help but wonder if this giant wasn't also a victim of plastic pollution.

Later back on the boat, I see a harem of sea lions vocalizing in the distance. They swim in pirouettes, flapping fins out of the water as if to wave as they slide by. On the shore, they crowd together by the dozen, lying around in the sunshine. I notice one of the smaller animals toward the back isn't moving; zooming in with my binoculars, I notice what looks like a fishing line wrapped around the torso of the sea lion corpse, another victim of entanglement.

In addition to orphaned fishing gear, plastic is a part of the fishing process for most. Fishers commonly use plastic bags as bait to hook crabs, a popular catch in Kino Bay. In the cleanup of Laguna la Cruz, the most common item found was plastic bags used for this purpose. In addition to cleanups, conducting awareness campaigns for fishers to properly dispose of fishing equipment and using bait and tackles instead of plastic for bait can be part of the solution. Reducing the use of single-use plastics by utilizing reusable containers is essential. For example, when shopping for groceries, I always bring a tote bag to reduce the need for plastic bags. Reducing plastic use, combined with organizing awareness and outreach campaigns in the community, gives me hope for a future with less plastic harming marine wildlife.

*Jake Krauss is a conservation biologist and science communicator dedicated to the protection of endangered species. He has worked as a wildlife biologist on conservation projects in Australia, Ghana, Ecuador, Peru, and Madagascar. Currently, he works in conservation communications for Prescott College's Kino Bay Center in Sonora, Mexico.*

## What Is the True Cost of Microplastic Pollution on Our Health?
*Kara Rogers*

If you pause and take a moment to look around you, wherever you happen to be, chances are you will see an object that contains or is made entirely of plastic. If you are outdoors, there is a high likelihood that the soil you are standing on and the air you are breathing contain particles of microplastics. If you are eating, there is even a risk that your food and the water you are drinking contain small amounts of microplastics.

Microplastics are everywhere. There is no escaping them. They are in countless products used by people every day, from plastic bottles and bags to packaging and shipping materials to synthetic clothing and personal care products. They are in the remote reaches of the high Arctic, where scientists have recovered tens of thousands of microplastics particles from a single liter of snow (Meyer 2019). And they are in wildlife.

Crustaceans that inhabit some of the deepest ocean environments in the world, like the Mariana Trench, are contaminated with microplastics (Jamieson et al. 2019).

All this begs the question, If microplastics are in soil, air, and water and in products we touch and use every day, does that mean microplastics are inside us too?

The unfortunate answer is yes. Scientists know microplastics are in humans because they have been recovered in human feces (Schwabl 2019). Other researchers have found that through air, food, and liquids, the average American ingests between 74,000 and 121,000 microplastics particles each year (Cox 2019). These numbers likely are underestimates.

But what exactly are microplastics? Technically, they are defined as small bits of plastic, fewer than 5 mm in length, which is generally comparable to the size of a sesame seed ("What Are Microplastics?" 2020). Chemically, they are made up of chains of carbon and hydrogen atoms. They often contain added chemicals, such as bisphenol A (BPA), dioxin, persistent organic pollutants, phthalates, and polychlorinated biphenyls, which can leach out of plastics and enter the environment ("ToxTown" 2020).

Microplastics from cosmetics, exfoliants, toothpaste, and other products get washed down the drain and sent into wastewater systems. The laundering of synthetic clothing also sends microplastics into wastewater. The particles are so tiny that they pass directly through filtering systems that otherwise catch contaminants. In this way, microplastics end up in our rivers, lakes, and oceans ("What Are Microplastics?" 2020; "ToxTown" 2020). Particles of microplastics, including those shed through the weathering and breakdown of larger plastics, can enter in the air and settle onto foods, plants, and soil, where humans and other animals come into contact with them (Cox 2019; Meyer 2019).

But even while microplastics are so pervasive, little is known about how they may be impacting our health. Microplastics in marine ecosystems potentially can move through the marine food web and ultimately contaminate sources of seafood ("Tox-Town" 2020). But what happens when we eat them?

The short answer is that experts do not yet know. Researchers with the World Health Organization note that microplastic particles more than 150 micrometres in size probably are not absorbed by the human body, and even in the case of smaller particles, absorption probably is minimal ("Microplastic Pollution Is Everywhere, but Not Necessarily a Risk to Human Health" 2019).

There is concern, however, that microplastics can cause inflammation in tissues and cells, potentially stimulating prolonged or abnormal inflammatory responses (Smith 2018). On a cellular level, preliminary evidence indicates that microplastics are harmful. Under laboratory conditions, for example, researchers have found that some 60 percent of human immune cells that come into contact with microplastics die. This far surpasses the rate of cell death for immune cells that deal with foreign bacteria in our bodies (Spink 2019). Accelerated immune cell death and chronic or abnormal inflammation could have serious implications for human health.

Also concerning are the chemical additives in microplastics. BPA, dioxin, and phthalates are toxic to the endocrine system, which oversees basic biological processes, such as metabolism and reproduction. Dioxin is a known carcinogen (cancer-causing substance) in humans. BPA is a possible carcinogen. Nonetheless, how much of these additives in microplastics actually make it inside us is unclear.

Even with the red flag of toxic chemical exposure from additives in microplastics, plastic production continues to increase globally, fueled by modern "throwaway" lifestyles. From 1950, when plastic production experienced its first rapid phase of growth, to 2015, some 381 million tonnes of plastic were produced. From 1980 to 2015, only one-fifth of all global plastic waste was recycled, while more than half was discarded as trash and one-quarter was incinerated (Ritchie and Roser 2018).

With so much plastic making its way into landfills and countless microplastics particles being washed into our oceans, the environment is being polluted with plastics at an unprecedented rate. And the situation will only continue to get worse if we do not soon abandon our addiction to plastics. As a consequence, the true cost of our plastic habit to the environment and our own health is yet to be realized.

## References

Cox, Kieran D., et al., 2019. "Human Consumption of Microplastics." *Environmental Science & Technology*. 53(12): 7068–7074. https://pubs.acs.org/doi/10.1021/acs.est.9b01517.

Jamieson, A. J., et al. 2019. "Microplastics and Synthetic Particles Ingested by Deep-sea Amphipods in Six of the Deepest Marine Ecosystems on Earth." *Royal Society. Open Science*. 6: https://doi.org/10.1098/rsos.180667.

Meyer, Robinson. 2019. "A Worrisome Discovery in High Arctic Snowfall." *The Atlantic*. https://www.theatlantic.com/science/archive/2019/08/microplastic-air-pollution-real/596119/.

"Microplastic Pollution Is Everywhere, but Not Necessarily a Risk to Human Health." 2019. United Nations, UN News. https://news.un.org/en/story/2019/08/1044661.

Ritchie, Hannah, and Max Roser. 2018. "Plastic Pollution. Our World in Data." https://ourworldindata.org/plastic-pollution.

Schwabl, Philipp, et al. 2019. "Detection of Various Microplastics in Human Stool: A Prospective Case Series." *Annals of Internal Medicine*. 171(7): 453–457. https://annals.org/aim/article-abstract/2749504/detection-various-microplastics-human-stool-prospective-case-series.

Smith, Madeline, et al. 2018. "Microplastics in Seafood and the Implications for Human Health." *Current*

*Environmental Health Reports.* 5(3): 375–386. https://www
.ncbi.nlm.nih.gov/pmc/articles/PMC6132564/.

Spink, Abigail. 2019. "New Evidence Points to Microplastics'
Toxic Impact on the Human Body." Geographical.
https://geographical.co.uk/people/development/item
/3422-microplastic-human-cells.

"ToxTown." 2020. "Microplastics." NIH U.S. National
Library of Medicine. https://toxtown.nlm.nih.gov
/sources-of-exposure/microplastics.

"What Are Microplastics?" 2020. National Ocean Service.
https://oceanservice.noaa.gov/facts/microplastics.html.

*Kara Rogers is a science writer and editor specializing in biomedi-
cine and the life sciences. She is the author of* The Quiet Extinction: Stories of North America's Rare and Threatened Plants
*(University of Arizona Press, 2015) and* Out of Nature: Why
Drugs from Plants Matter to the Future of Humanity *(University of Arizona Press, 2012). She is a member of the National
Association of Science Writers.*

## Small Town Makes Big Impact
*Diana Reynolds Roome*

"Wouldn't it be awesome if Talent became the first city in Oregon to ban plastics?" The small tousle-haired figure speaking
into a microphone before Talent City Council was Zoë Wil-
liams, 10-year-old author of a children's book about a shark
whose life changes dramatically when she inadvertently swal-
lows a plastic bag. Zoë was one of several speakers that April
evening, arguing that it was time to say goodbye once and for
all to the damaging convenience of throwaway plastic. Her tes-
timony garnered applause from council members and audience.

Less than a year later, Zoë got her wish. On January 1, 2020,
Talent, a town of little more than 6,000 nestled between mountains in the Rogue Valley, Oregon, introduced an ordinance

banning disposable single-use plastic at all its food businesses—restaurants, cafés, food trucks, and supermarkets. Though there were only about 25 of those in town, it took years of planning, research, and communication to put the ban into effect. Back in 2010, Sharon Anderson, who has lived in Talent for 20 years, took a Master Recyclers course that made her acutely aware of the massive problem of plastic thoughtlessly thrown away every day.

"I realized there is no 'away,'" said Sharon. That's because recycling only takes care of a small fraction of the problem. Less than 9 percent of plastic is recycled in the United States. The vast majority is tossed into landfills where it disintegrates over decades and washes into streams, then into the ocean. There plastic does massive damage, as Zoë's book describes. It never goes away completely. Oregon's long, beautiful coastline has been pristine for most of its history. It's hard to bear the thought of the ocean and beaches full of plastic debris.

"Back when I became a Master Recycler, we were each asked what we planned to do with our new knowledge about waste," said Sharon. "I thought, if any place can change for the better, Talent can." A committed environmentalist, she was already busy working on several volunteer projects, including picking up litter, planting trees, and collaborating on a clean energy plan for the town. But each year, more information came out about the ways plastic hurts our environment. The time had come to tackle the problem.

Getting together with a few others who felt strongly about waste, she formed a volunteer Zero Waste Team (TZWT) as a subcommittee of the City's Together for Talent Committee, to decide what could be done. At first it seemed daunting to take on an issue that affects the whole world, but anyone can start close to home. Along with despair about damage done by plastics came a strong spirit of determination.

Challenges are everywhere. Changing people's habits is never easy. Everyone is used to convenience—the ease of picking up prepared food, eating on the run and throwing away the container. Buy a drink, take a plastic cup and straw—no big deal, right? Yet disposable plastic is a problem that grows and compounds every day. Recycling cannot keep up with the problem, so we have to get more fundamental. We must say no to those plastic takeaway items and yes to reusable containers and utensils that we can probably scrounge up at home.

The Talent Zero Waste Team started by informing people, widening perspectives, and inviting ideas. At the local Harvest Festival, they provided new options for all kinds of recycling, with a town-wide Recycle Drop Off for light bulbs, appliances, and electronic items. Discussions followed, about how to be less wasteful of all resources. Movies showed in heartbreaking detail the damage done by plastic to birds, animals, and humans and how items made from fossil fuels hurt the soil, water, air, and nature itself. Older members of the community remembered a day when plastics did not exist, and everyone thrived without it. Younger members became acutely aware of ways our hurried lives and need for convenience amounted to a daily assault on the environment.

On January 1, 2020, eight months after that April meeting, when Sharon, Zoë, and a small crowd of anti-wasters brought their ideas before the City Council, a quiet revolution took place in Southern Oregon. Talent's cafés, restaurants, food trucks, and other food and beverage vendors started to phase out all types of disposable plastic for eat-in, takeout, and left-over foods. This included stirrers, utensils, containers, plates, and bowls, both plastic and bioplastic (which must often be broken down in an industrial composter). Food vendors knew that changing habits and buying practices would require time, effort, and some extra costs. They realized some customers would be unhappy. Though alternative materials are constantly improving, there's still a long way to go before they satisfy everyone.

If Talent was the first city in Oregon to implement a comprehensive ban on disposable plastics, those food businesses went one step better. The key to reducing waste is reusing. So several offered a discount for customers who brought their own container or cup. Others used wide-mouthed glass jars for leftovers. Some sold sturdy hot-and-cold cups with their logo. For a small charge, one offered durable food containers that could be exchanged for a clean one on each visit. The message was simple, inexpensive, and old fashioned: Bring Your Own!

"It's all about changing habits," said Sharon Anderson. "Talent wants to go further than other cities and reduce the use of any single-use food service product regardless of how 'green' its material may seem to be. Hence, our campaign to get folks to reuse their own items. It's a challenge—but we must do it. Refuse and reuse!"

In January 2020, Talent's Zero Waste Team held a kickoff party at a popular local pizzeria to introduce and celebrate the idea of Bring Your Own to the community. Zoë was there, signing her book, *Coral's Quest*. So were TZWT volunteers who had sewed striking tote bags with the Zero Waste logo, as well as BYO pouches containing knife, fork, spoon (or spork), metal straw, and napkin. One, Lynne Likens, had devised a giant silver fork with the slogan, "May the fork be with you!" A Trash Fashion contest brought in people sporting bizarrely inventive costumes created from bubble wrap, cardboard boxes, plastic forks, and polystyrene peanuts.

"We're clear that this is a priority," said Mayor Darby Ayers-Flood, smiling as she surveyed the crowd.

Neighbors Zoë and Sharon shared a warm hug of triumph. Between them, spanning seven decades, their joint commitment to say goodbye to throwaway plastic was helping Talent move into a plastic-free future.

Since then, the Covid-19 pandemic reversed many of these gains, as food businesses were no longer allowed to handle BYO containers. The effort to reduce plastic use may have slowed down, but a new impetus arrived a few months after the Talent plastics ban went into effect when Zoë was declared a semifinalist in U.S. Cellular's Future of Good program. In a surprise celebratory parade, she was presented with a giant mock check for $10,000, representing real funds to help further her goals to reduce plastics. Through the Talent Zero Waste team, Zoë aims to use the money to work with local schools to eliminate plastic utensils and replace them with all reusable ones.

The young author also has another book in mind, about Coral's efforts to help clean up earth and oceans. She sees writing as a way to reach people she can't meet in person. "Like Coral says in my book, 'It only takes one person to make a difference.'"

*Author of children's books and articles on health, travel, and the arts, Diana Reynolds Roome has lived in England, Nepal, California, and now southern Oregon, where reducing pollution of all kinds is a priority.*

## Environmental Hazards of Bottled Water
*Ellen Rubin*

It looks so harmless, even helpful, yet that bottle of water in your hand is destroying the planet.

While people first bottled water in the 1700s, it was a specialty item confined to therapeutic drinking or bathing water, and people were subject to questionably safe water supplies for everyday use. However, with the advent of chlorinating municipal water supplies in the early twentieth century, safe drinking water became readily available to the general public. There is no safety reason to buy bottled water.

With the decline in carbonated soft drink sales in the 1990s, manufacturers began looking for an alternative source of revenue. They expanded on the niche market that waters such as Perrier had created in the 1970s, marketing bottled water as a status symbol. The modern bottled water industry emerged in the 1990s. Two factors contributed to its emergence: thousands of people were sickened by parasites in Milwaukee's municipal water system and new plastic technology led to a decrease in the cost per bottle.

## Low-Cost, Clear Plastic

Water was historically sold in glass bottles or fairly thick, non-clear plastic. Polyethylene terephthalate (PET) was first invented by DuPont chemist Nathaniel C. Wyeth in the 1970s. It is a form of polyester. It wasn't until 1993 that PET was used to create a thinner, lighter, less expensive bottle that was more appealing to consumers and more cost-effective. While Coca Cola and Pepsi were some of the first proponents of bottled water with their Dasani and Aquafina brands, water was still fairly expensive. This changed when the manufacturer, Niagara, focused on technological and automation innovations to reduce costs so water could be attractively priced for the average person. Niagara markets its water under a number of generic and brand names. Among their innovations was to shrink wrap cases without cardboard and reduce the weight of a half-liter bottle from 23 grams to 7 grams, which reduced manufacturing and transportation costs (Moss 2018). The goal was to lower the price enough to change water from a luxury item to a convenience item. Today, Niagara sells two-thirds of all bottled water sold in the United States. With a lower cost, demand increased to the point that 1 million bottles of water are sold every minute.

Among the issues that arose from the explosion of bottled water sales are (1) environmental cost of producing the plastic for the bottle, (2) the bottling process and transportation, (3) pollution caused by the bottles themselves, and (4) the economic and environmental impact of bottling plants.

## Water Bottle Production

Petroleum products are used to create PET. Producing a single plastic bottle requires one-quarter of its volume in oil. Currently, over 17 million barrels, or 47 million gallons, of oil is needed to manufacture the plastic for water bottles (Goldschein 2011). This is the same amount used to power 190,000 homes for a full year. The manufacturing process produces 2.5 million tons of carbon dioxide. In addition to oil, creating the bottle uses twice as much water as it will eventually contain (Richardson 2016). Transportation costs are incurred at each stage, including: shipping oil to plastic manufacturers, the plastic to the bottling manufacturer, the bottles to the water plant, and then the bottled water to the retailer (one-quarter of bottled water is not sold regionally) (Printwand 2019).

## Bottling Process

The purification process also uses twice the volume of water that the bottle will contain; therefore, the quart bottle you take off the shelf uses three to four quarts of water in its production. Water bottling is hugely inefficient, waterwise. Further, 90 percent of the price you pay is profit for the company. The bottle costs a few cents to manufacture, and the water it contains costs a fraction of a cent.

## Environmental Impact

PET bottles are not readily biodegradable or recyclable. Very few people even attempt to recycle their water bottles. In 2005, of the 30 billion bottles sold, only 9 percent were recycled (Nace 2017). Instead, the bottles are incinerated, which causes toxic fumes; 80 percent are dumped in landfills, where plastic chemicals may leach into the water table; or they are thrown away as litter, where they can be consumed by wildlife, often with fatal consequences (Richardson 2016). Recycled PET plastic can't be used for any type of food or water product because of the chemical toxicity of the plastic from diethylhydroxylamine (DEHA) and benzyl butyl phthalate (BBP). Water bottle caps are not recyclable at all. If not recycled, PET bottles take 450 years to biodegrade (Lake 2015). Of the 1,500 bottles of water consumed every second, or 250 billion pounds per year (371 pounds per person), only a fraction is actually recycled.

Plastic litter has become an urgent problem. It has been estimated that by 2050 there will be more plastic in the ocean, by weight, than fish. Not only does the plastic cause damage to the ecosystem, but it is also ingested by sea life. Increasingly, there is plastic in the seafood that we consume. Ghent University in Belgium found that the average person ingests up to 11,000 pieces of plastic yearly (Nace 2017). Plastic is also found in lakes and the ocean. Around 22 million pounds of plastic ends up in the Great Lakes every year. Once it breaks down into microplastic and is eaten by fish, it works its way up the food chain to humans (Chow 2017). Microplastic bits have also been found throughout the land, even in the upper reaches of the Rocky Mountains.

## Economic and Human Impact of Factories

Bottling factories are not necessarily a boon to local economies. Many are located in areas, such as California, that don't have an overabundance of water available. Nestlé actually created a water shortage in Pakistan when they opened a water bottling plant there, as well as one when they located a plant in California. To avoid similar environmental impact, Washington state has recently proposed legislation that would make it illegal to take water from local sources for bottled water. Other detrimental effects include: damaging local wildlife and fish populations, raising water prices for local residents, and increasing road repair costs. Plants have minimal impact on employment rates, have a higher incidence of illness and injury, and provide only low-paying jobs. Economic benefits from a plant are limited to the headquarters and the highly paid executives (Food & Water Watch 2008).

Bottled water does play a vital role when there are natural disasters such as floods, hurricanes, or tornadoes that contaminate the local water supply, but the ecological and economic toll that widespread reliance on bottled, rather than tap, water has is substantial. Very few people, outside the owners of the companies, benefit from bottled water, especially when municipal sources provide a safe, very inexpensive alternative in water from your faucet.

## References

Chow, Lorraine. 2017. "1 Million Plastic Bottles Bought Every Minute, That's Nearly 20,000 Every Second." EcoWatch. https://www.ecowatch.com/plastic-bottle-crisis-2450299465.html.

Food & Water Watch. 2008. "The Unbottled Truth about Bottled Water Jobs." https://www.foodandwaterwatch.org/insight/unbottled-truth-about-bottled-water-jobs.

Goldschein, Eric. 2011. "15 Outrageous Facts about The Bottled Water Industry." Business Insider. https://www.businessinsider.com/facts-bottled-water-industry-2011-10.

Lake, Rebecca. 2015. "Bottled Water Statistics: 23 Outrageous Facts." Creditdonkey. https://www.creditdonkey.com/bottled-water-statstics.html.

Moss, Robert. 2018. "How Bottled Water Became America's Most Popular Beverage." Serious Eats. https://www.seriouseats.com/2017/07/how-bottled-water-became-americas-most-popular-beverage.html.

Nace, Travor. 2017. "We're Now At A Million Plastic Bottles Per Minute—91% of Which Are Not Recycled." Forbes. https://www.forbes.com/sites/trevornace/2017/07/26/million-plastic-bottles-minute-91-not-recycled/.

Printwand. 2019. "Stop Polluting the Planet with Disposable Plastic Bottles." https://www.printwand.com/blog/plastic-bottle-pollution-effects-facts.

Richardson, Jake. 2016. "21 Facts about Bottled Water, The Environment, & Human Health." Insteading. https://insteading.com/blog/21-facts-bottled-water-environment-human-health/.

*Ellen Rubin is a lifelong learner with degrees in history, library science, and law. In her spare time, she enjoys reading, doing fiber arts, gardening, and baking—with chocolate!*

## The Reckoning
### Das Soligo

The many benefits of plastics are very well known and documented. In terms of convenience, recreation, logistical support, and protection of health and safety, there are very few alternatives that can tick these boxes so cost-effectively and easily when a designer is considering a package or material type for their product.

The unique attributes of plastics have led to their widespread use and made their presence ubiquitous in nature. Plastic litter is easily visible in virtually any location on the planet, in forests, deserts, and in waterbodies of all types. A recent report estimates that by 2050 there will be more plastic in the ocean than fish, as the equivalent of a garbage truck of plastic is finding its way to the ocean every minute of every day (Ellen MacArthur Foundation 2017, 13).

Unfortunately, plastics are just not captured adequately to reutilize the resources they contain. Besides the litter issue, the nature of plastic's chemistry makes it challenging to effectively recycle the varied number of plastics with a great degree of success. Plastics are highly recyclable materials, but products and packaging come in all shapes, sizes, and forms and sometimes contain mixes of a number of different plastic types.

While the majority of plastic packaging and products acceptable in municipal recycling programs are able to be recycled, recycling facilities cannot handle and process many of the other plastic materials that come through these plants. It is estimated that only 9 percent of the plastic waste ever produced has been recycled (United Nations Environment Programme 2018, 6). This is in large part due to the many challenges of recycling the vast variety of plastic material, coupled with a global culture that views plastics as cheap and highly disposable.

In general, there is very little regard for purchasing mass-produced junk; products and packaging that are used briefly and disposed of remain in a landfill, or the environment, for potentially thousands of years. In my career in the waste management field, I observe the tons of plastic waste that is piling up in our landfills, and I wonder what archaeologists in the future will think of when exploring our communities and waste sites. Surely our habits and throwaway culture will not reflect well on us, as the abundance of resources contained in plastic waste has been willfully discarded and unused. To make matters worse, we clearly know the error in our ways.

As a young man studying environmental studies, climate change, and waste management 20 years ago, I was very angry and frustrated by society's disregard for the environment. I felt somewhat hopeless, as we clearly had the knowledge to know that our practices are wasteful and unsustainable, and they are taking a massive toll on living organisms. Yet these practices were continuing in the face of political expediency and inertia. There appeared to be little likelihood that this would change any time in the near future.

As my sense of despair increased, a wise man told me that anyone can be angry at the world and point out all of its problems, but only true leaders will come up with solutions to these problems.

From that time on, I made a conscious choice to reframe and refocus my frustration to use it as motivation. I began to understand and appreciate that many small positive actions taken and repeated locally and around the world will help to create a more sustainable future. In the past 20 years, I have witnessed the mainstreaming of environmental movements. People of all ages, and especially the youth of today, are demanding a more sustainable economy. In politics and in conversations around the world, the topic of plastic waste has been debated and discussed to an extent not seen before.

As of February 2019, 112 cities, regions, and countries around the world have implemented some sort of restriction or ban on single-use plastics (Telesetsky 2019), an astounding figure that is undoubtedly higher now, as these regulatory actions are increasingly more common. At first glance, a ban on straws or plastic bags seems to be a symbolic drop in the ocean and a somewhat insignificant action in the face of an enormous global challenge. However, these actions fit a pattern of growing awareness around environmental issues. The single-use plastics restrictions and bans signal that society is demanding change around this complex issue. There is an emerging willingness to pay for more sustainable alternatives to single-use plastics, while forgoing their convenience for the betterment of the environment.

This evolution in environmental consciousness is starting a dialogue around what a more sustainable economy would look like. Concepts like true cost accounting, ecosystem services, extended producer responsibility, and transitioning to a circular economy are coming to the forefront in considering sustainability. As the magic and wonder of nature is increasingly understood, the field of biomimicry is working to imitate nature's elegance in its systems, elements, and models, in order to overcome complex human problems.

In these and other concepts, there will be so many opportunities to reduce wasteful practices. Alternatives to disposable plastics are here and are continually developing. Innovation and regulation can help ensure that plastics that need to be disposed of are able to be captured and reintegrated into new products and packaging in ways that beneficially and fully utilize the resources that plastics contain.

There is truly hope for a more sustainable future. It has been decades in the making, but that vision is arriving more quickly than it once looked like. To help usher in a better tomorrow, we need all the true leaders of today to continue to develop solutions to the many environmental issues that plague our people and planet. We need each one of us to make the choice to take actions, whether small or large, to help facilitate the societal evolution and progression of our species.

## References

Ellen MacArthur Foundation. 2017. "The New Plastics Economy: Rethinking the Future of Plastics and Catalyzing Action." https://www.ellenmacarthurfoundation.org/assets

/downloads/publications/NPEC-Hybrid_English_22-11
-17_Digital.pdf.

Telesetsky, Anastasia. 2019. "Why Stop at Plastic Bags and
Straws? The Case for a Global Treaty Banning Most Single-
use Plastics." Public Radio International (PRI). https://
www.pri.org/stories/2019-02-07/why-stop-plastic-bags-and
-straws-case-global-treaty-banning-most-single-use.

United Nations Environment Programme. 2018. "Single-
Use Plastics: A Roadmap for Sustainability." https://
wedocs.unep.org/bitstream/handle/20.500.11822/25496
/singleUsePlastic_sustainability.pdf.

*Das Soligo is the Manager of Solid Waste Services at the County
of Wellington. He has worked his entire career in the county after
graduating with distinction from the University of Toronto, studying Geography—Resource Management and Environmental
Studies. Das is passionate about the environment and the waste
industry and is an active participant in a number of related com-
mittees, initiatives, and organizations.*

# 4 Profiles

This story of plastics and microplastics, to a large extent, is the story of individuals and organizations that have made contributions to the field in the past and at the present time. The list of such individuals and organizations is very long, and the specific names listed here can be no more than illustrative of those who have contributed to the rise of the plastics industry, its contributions to society, and solutions that have been devised and proposed for dealing with today's problems associated with plastic use. For a much longer and more detailed summary of the work of more than 400 individuals in the field, also see Plastics Hall of Fame (https://www.plasticshof.org/).

## Alliance to End Plastic Waste

The Alliance to End Plastic Waste was created in January 2019 by a group of companies involved in the production, processing, or use of plastic materials. Among the current members are BASF, Chevron Phillips, Dow, Exxon Mobil, Formosa Plastics, Mitsubishi Chemical Holdings, Mitsui Chemicals, P&G, Royal Dutch Shell, Sumitomo Chemical, and Total. The primary goal of the organization is "to end the flow of plastic waste into the environment" (https://endplasticwaste .org/). The organization has pledged a total of $1.5 billion over the next five years to achieve this objective. It proposes using

For over 50 years, Greenpeace has been one of the premier organizations concerned about the environmental protection of the oceans, including that arising from plastic pollution. (Fabrizio Robba/Dreamstime.com)

the technical skills possessed by its thousands of employees to devise elements of a possible solution to the problem and find ways of implementing those discoveries in everyday life.

The Alliance has built its program around four major elements: infrastructure, innovation, education, and cleanup. The infrastructure element of this effort is based on the fact that while there has been a huge growth in the production and use of plastics, there has not been a comparable increase in the facilities needed to deal with plastic wastes. The organization plans to work with local communities to develop solid waste management programs that can more efficiently collect and safely dispose of plastic wastes produced in those communities. The greatest effort in this aspect of the program will be directed at developing countries, where even the most basic solid waste management techniques are not well understood or economically possible.

As the name suggests, the innovation aspect of the program is designed to support and encourage research on technologies that can be used to minimize wastes in the first place or recycle and recover wastes for future reuse. The education component of the Alliance program is intended to work with communities, private businesses, and governmental agencies at all levels to improve understanding of the plastic waste problem and ways of dealing with that problem in those specific situations. Finally, the cleanup feature of the program takes into account that one major feature of the plastic waste problem is the existence of sites where the problem already exists and requires expert advice and attention for their resolution.

The Alliance is still only a year old at the time of this writing, so more detailed information about its work should become available in the near future. Its home page online is https://endplasticwaste.org/.

## Leo Baekeland (1863–1944)

Baekeland is arguably the most famous name in the history plastics. In 1906, he invented the first entirely synthetic plastic, a substance that was inexpensive to make, sturdy, machinable, and resistant to environmental damage. For this invention, Baekeland is often called the Father of the Plastics Industry.

Leo Henricus (Hendrik) Arthur Baekeland was born in Ghent, Belgium, on November 14, 1863. His parents were Charles Baekeland, a cobbler, and Rosalie (Merchie) Baekeland, a housemaid. He attended elementary school in Ghent before matriculating at the Ghent Municipal Technical School. Upon his graduation from high school, he was awarded a scholarship from the city of Ghent that allowed him to begin his studies in chemistry at the University of Ghent in 1880. He earned his bachelor's degree from Ghent in 1882 and then stayed on to study for his PhD, which he received *maxima cum laude* two years later. He taught briefly at the Government Higher Normal School in Bruges from 1887 to 1889, before returning to Ghent as associate professor of chemistry.

In 1889, Baekeland was awarded a travel scholarship that allowed him to travel overseas to continue his studies. He left with his new wife to visit the United States, where he hoped to pursue his studies in photography, a topic in which he had been interested since his youth. Baekeland had already made one new improvement in photographic technology at Ghent in 1887. That invention provided a method for developing photographic solutions made with water rather than the more complex system using dry materials. In New York, Baekeland first took a job with E. and H. T. Anthony Photographic Company to continue this line of research. Two years later, he decided to branch out and become a consulting chemist. When that business failed, he returned to the Anthony company, where he made one of the two inventions for which he is most famous, Velox photographic paper.

Velox was the first material that could be used to develop photographs under artificial light, rather than natural light. The invention revolutionized the practice of commercial photography, and Baekeland was able to sell the product to the Eastman Kodak Company for $750,000 in 1899 (equivalent to more than $21 million in today's currency). That sale essentially made Baekeland financially independent for the rest of his life. As he later said, "Thus at thirty-five I found myself in comfortable financial circumstances, a free man, ready to devote myself again to my favorite studies. Then truly began the very happiest period of my life" (Charles F. Kettering. 1946. "Biographical Memoir of Leo Hendrik Baekeland." Biographical Memoirs. National Academy of Sciences, 286. http://www.nasonline.org/publications /biographical-memoirs/memoir-pdfs/baekeland-leo-h.pdf).

With part of the proceeds from his sale of Velox, Baekeland purchased a large home in Yonkers, New York, called Snug Rock. He installed a sophisticated chemical laboratory there and, in 1907, invented a new plastic that he named after himself, Bakelite. The discovery occurred while Baekeland was working on a substitute for the natural product, shellac. That field of research was an active one at the time because natural sources of the substance had begun to disappear. And shellac was a critical component in many industrial and commercial operations. Baekeland's work in this area did not produce successful results, but along the way, he discovered a mechanism for making a new product that eventually turned out to be far more important and useful than shellac. He produced the substance by reacting two organic compounds, phenol and formaldehyde, with each other. He discovered that the conditions under which the reaction was carried out were crucial to the physical properties of the material produced. Baekeland received a patent for his invention in 1909 and announced the discovery at a meeting of the New York Section of the American Chemical Society in the same year. A year later, he formed the General Bakelite Corporation, a business that he later sold to Union Carbide in 1939. Bakelite is still used today in one form or another in the manufacture of a variety of objects, including jewelry, toys, radio and telephone casings, electrical insulators, toilet seats, dice, and kitchenware.

The discovery of Bakelite enriched Baekeland even more, and he was largely able to pursue his own research interests throughout the rest of his working career. In 1917, he was appointed honorary professor in Columbia University's new Department of Chemistry, a position in which he served with considerable distinction for many years. He received many honors during his lifetime, including honorary doctorates from the universities of Brussels, Columbia, Edinburgh, and Pittsburgh, as well as the Nichols Medal of the New York Section American Chemical Society, John Scott Medal of the Franklin Institute, Willard Gibbs Medal of the Chicago Section of the American Chemical Society, Chandler Medal of Columbia University, Parkin Medal and Messel Medal of the Society of the Chemical Industry of London, Franklin Medal of the Franklin Institute, Grand Prize of the Panama-Pacific Exposition of 1915, Pioneer Trophy of the Chemical Foundation, Scroll of Honor of the National Institute of Immigrant Welfare, Officer of the French Legion of Honor, Officer of the Order of the Crown (Belgium), and Commander of the Order of Leopold (Belgium). Baekeland died in Beacon, New York, on February 23, 1944, and was buried in Sleepy Hollow Cemetery in Sleepy Hollow, New York.

## Wallace Carothers (1896–1937)

In his biographical sketch on the Plastics Hall of Fame website, Carothers is described as follows: "Few individuals have had such a profound impact on materials science and industry as Carothers" (https://www.plasticshof.org/members/wallace -h-corothers). The E.I. DuPont de Nemours & Company, where Carothers worked from 1896 to 1937, called him "one of the most brilliant organic chemists" the company had ever employed (https://lemelson.mit.edu/resources/wallace -hume-carothers). To professional chemists, he may be best remembered for his research on polyamides (https://www.ias .ac.in/article/fulltext/reso/022/04/0339-0353). And to the average person, his name will always be associated with the invention of nylon and the beginning of the first synthetic textile industry (https://www.acs.org/content/acs/en/education/whatischemistry/landmarks/carotherspolymers.html).

Wallace Hume Carothers was born in Burlington, Iowa, on April 27, 1896. His father, Ira Hume Carothers, spent his life as an educator, serving as teacher and vice president of the Capital City Commercial College, in Des Moines, for 45 years. His mother, Mary Evelina McMullin Carothers, was a housewife originally from Burlington, Iowa. He attended public schools in Des Moines, where his family had moved when Wallace was five years of age. Upon his graduation from North High School in Des Moines, he enrolled at the college where his father was employed, earning a degree in accounting and secretarial curriculum in only a year. He then continued his studies at Tarkio College, in Tarkio, Missouri. He began a major in chemistry there, while simultaneously working as an assistant instructor in the college's commercial department. He then continued his bifurcated experience at Tarkio, taking a teaching position in the English department, while completing his degree in chemistry.

After completing his BS in chemistry at Tarkio in 1920, he enrolled in the master's program in chemistry at the University of Illinois, where he was awarded his MS in chemistry a year later. He then took a year off to teach chemistry at the University of South Dakota, before returning to Illinois, where he received his PhD in chemistry in 1922. He remained at Illinois as an instructor in chemistry until moving on to Harvard University in 1926 as an instructor in organic chemistry. After a single year at Harvard, Carothers was offered a job at the DuPont chemical company, which had just created a division for pure research in chemistry. After much soul-searching, Carothers accepted the offer, at least partly because of the substantial increase in salary, from $267 per month at Harvard to $500 per month at DuPont.

Progress in research at DuPont was slow at first, but by 1930, the first results had begun to appear. The most important of these was a new polymer called *neoprene*, produced by the polymerization of the newly synthesized monomer, *chloroprene*. The new polymer soon found use as the first synthetic material capable of being used in the manufacture of tires. Given the glimmers of a new world war with the likely consequence of loss of most of the world's natural supply of rubber, the discovery of neoprene was of supreme importance not only to the field of synthetic chemistry but also to the tire industry itself.

By 1934, Carothers and his team had turned their attention to an entirely new type of polymer, polyamides. Polyamides are long-chain compounds in which monomers are linked by amide bonds. They have the property of forming very large linear molecules capable of being made into fibers. The most famous of these early compounds was named *nylon-6*, a substance that was to become the parent compound of a whole family of similar compounds, the *nylons*. Carothers's research provided a model on which similar types of plastics could be formed, making available a whole new category of materials from which a variety of fabrics can be made. In the United States alone in 2018, more than 590,000 metric tons of nylon were produced.

Carothers's apparent enormous success with polyamides and nylons proved to be of little help in aiding him to overcome his own self-doubt. From his earliest college days, he felt and expressed a deep unhappiness with his life, extending to the point that he had a reputation at Illinois as always carrying a cyanide pill with him. At one point during his years at DuPont, he wrote to a friend that "I go through at least a dozen violent storms of despair every day." By the mid-1930s, he was seeing a psychiatrist on a regular basis to get help for what was later diagnosed as manic-depressive disorder. In January 1937, he beloved sister Isobel died of pneumonia, an event that seemed to be the tipping point in his mental health. Four months later, he committed suicide by drinking a glass of lemon juice containing a cyanide pill. He had reached the point, he had said, where, in spite of his successes, he had not really accomplished anything of importance in his life and had run out of ideas for new research (https://www.sciencesource.com/archive/Wallace-Carothers--American-Organic-Chemist-SS2192188.html). His suicide occurred at a hotel in Philadelphia on April 29, 1937, when he was just 41 years old.

Carothers received almost no honors or awards during his lifetime, although a residence and dining hall at the University of Texas at Austin is now named in his honor.

## The 5 Gyres Institute

The 5 Gyres Institute was founded in 2009 by Anna Cummins and Marcus Eriksen. Cummins received her undergraduate degree in history from Stanford University and her masters in International Environmental Policy from the Middlebury Institute for International Studies. Ericksen earned his BS in earth science and his PhD in science education at the University of Southern California. The two met on an 88-day voyage from California to Hawaii aboard the Junk Raft, a sailing ship made entirely of waste plastic materials, including 15,000 plastic bottles. During the trip, Cummins and Ericksen (who became engaged on the voyage) discovered the second Great Pacific Garbage Patch and documented and quantified its composition of plastic materials. They also observed and recorded the impact of plastics on marine animals living in the area. The whole story of this experience can be found in Eriksen's 2017 book, *Junk Raft: An Ocean Voyage and a Rising Tide of Activism to Fight Plastic Pollution* (Boston: Beacon Press). A scholarly paper reporting the findings of the voyage is also available in the literature and online as "Plastic Pollution in the World's Oceans: More than 5 Trillion Plastic Pieces Weighing over 250,000 Tons Afloat at Sea" (*PLoS One.* 2014; 9(12): e111913. https://doi.org/10.1371/journal.pone.0111913).

The goal of 5 Gyres is "to empower action against the global health crisis of plastic pollution through science, education, and adventure." Over the 10-plus years of its existence, it has conducted 19 research voyages covering more than 50,000 miles, primarily across the North Pacific and South Pacific gyres of the Great Pacific Garbage Patch. Their work has been documented in more than two dozen published papers on topics such as "The Case for a Ban on Microplastics in Personal Care Products," "The Plastisphere—The Making of a Plasticized World," and "A Global Inventory of Small Floating Plastic Debris." Many of their papers deal with specific research topics, such as "Microplastic pollution in the surface waters of the Laurentian Great Lakes." (For a complete list of these papers, see https://www.5gyres.org/publications.)

An important element of the 5 Gyres program is its group of ambassadors, individuals with a special interest in learning about, teaching about, and working on issues of plastic pollution in the oceans. Currently more than 1,400 individuals, aged 8–80, in 45 American states and 67 countries make up the Ambassadors program. They receive information about the topic in monthly webinars, take part in programs and events sponsored by the organization, and have access to a private Facebook page run by 5 Gyres. Their responsibilities include speaking on plastic waste topics at live and online events, developing and promoting social media campaigns on plastic wastes, conducting scientific research in the field, and making their own videos on some topic related to plastic wastes.

Two specialized programs sponsored by 5 Gyres are a curriculum focusing on plastic wastes for elementary, middle, and high schools and a campaign known as Trashblitz. Trashblitz is designed to engage individuals in the effort to learn more about and act on issues of plastic waste at the personal and community levels. A description of four Trashblitz meetings held so far and planned for the near future can be found at https://www.5gyres.org/trashblitz.

The work of 5 Gyres is sponsored in part by partnerships with organizations who share and wish to promote its goals, such as Liquid Death Mountain Water, 4Ocean, Everything but Water, Natracare, Eagle Creek, Final Straw, LifeStraw, and

Saludos. In 2010, 5 Gyres became a founding member of the United Nations Global Programme on Marine Litter, and in 2017, it received recognition as having special consultative status with the United Nations Economic and Social Council. That designation is an honor accorded by a United Nations agency to nongovernmental organizations with expertise in some specific area of UN concern or action.

## Greenpeace

Greenpeace is one of the oldest, largest, best-known, and most successful environmental groups in the world today. It was founded in 1971 by Canadian activists, Dorothy and Irving Stowe. At the time, one of the most crucial environmental problems facing the world was nuclear power. After the military success of the first nuclear bombs at the end of World War II, governments rushed to further develop this powerful weapon of war. That development required testing, much of which was first conducted in the atmosphere. Eventually, these tests were moved underground in an effort to make them safer for the environment and humans. The actions conducted by the Stowe's and their friends in 1971 were aimed at preventing underground nuclear testing off the coast of Alaska, in a highly fragile tectonic zone. They obtained a rusty old fishing boat called the *Greenpeace* and set sail for the Alaska coastline. Their goal was not to do battle with the large navy ships responsible for conducting the tests, but simply to take a stand in their way. The effort was not successful, and the *Greenpeace* returned to Canada. Later, protest voyages were more successful, and nuclear testing in the area was discontinued. Equally important, news of the *Greenpeace* voyages had begun to spread worldwide, and a new environmental organization was born. No specific date or location is available for the formal creation of the organization. As one early member has written, "The truth is that Greenpeace was always a work in progress, not something definitively founded like a country or a company" (Patrick Moore. 2010. "Who Are the Founders

of Greenpeace?" Beatty Street Publishing. https://web.archive
.org/web/20121007000313/http://www.beattystreetpublishing
.com/who-are-the-founders-of-greenpeace-2/.).

The philosophy and program espoused by the original
Greenpeace group began to spread rapidly around the globe.
By 1979, as many as 20 such groups had formed in various
parts of the world. Each group adopted one or another environmental cause on which to focus: the slaughter of gray seals
on Orkney Islands, efforts to reduce or eliminate the harvesting
of whales, the dumping of radioactive wastes in the oceans, the
incineration of wastes on the open seas, the use of large drift
nets in commercial fishing, and even the use of chlorine-based
products in the manufacture of newsprint. At that point, individual groups began to appreciate the need and value of having
an international coordinating organization, and Greenpeace
International was formed. The moving force behind that action
was Canadian activist David McTaggart, who was able to bring
together existing Greenpeace groups into a single organization,
based in Amsterdam. McTaggart served as president of the new
group until 1991, although he remained active in its work until
his death in 2001.

Today, Greenpeace International (formally, Stichting Greenpeace Council) has a paid staff of more than 2,400 workers and
about 15,000 full-time volunteers. It consists of 26 national
and regional offices located in 55 countries. Some members
are Greenpeace Africa, Greenpeace Argentina, Greenpeace Belgium, Greenpeace India, Greenpeace Mediterranean, Green-
peace Nordic, and Greenpeace USA. The last of these groups,
Greenpeace USA, was formed by the consolidation of Greenpeace groups in San Francisco, Portland, Seattle, and Denver
in 1979. The individual groups of which Greenpeace International exists are largely independent and autonomous, focusing
on issues of special significance to each, with their own structure and modes of operation.

Both Greenpeace International and Greenpeace USA (as
well as other national chapters) are actively involved in efforts

to improve the public's awareness about plastic pollution, as well as solutions that have been suggested for the problem. Their websites carry special features on topics such as "Throwing Away the Future: How Companies Still Have It Wrong on Plastic Pollution 'Solutions,'" "Packaging away the Planet," "Going Plastic Free: The Rise of Zero Waste Shopping," "Recycling Isn't Dead, but Plastic Is Killing it," "The Making of a Plastic Monster," "Plastic Straws Are Just the Tip of the Iceberg," and "Envisioning the #ReuseRevolution."

## Walter Lincoln Hawkins (1911–1992)

Walter Hawkins is perhaps the best-known and most successful African American plastics researcher in history. His name does not appear in the Plastics Hall of Fame, which has no African American members, but nonetheless, he is widely honored for his inventions and discoveries in the field.

The most important of these inventions was the creation in the 1950s of a covering for underground telephone cables, such as those used to carry long-distance calls across the Atlantic Ocean. When those cables were first laid, they were coated with a lead-based material, which turned out to be highly undesirable because of the toxic properties of lead compounds. A substitute proposed for resolving this problem was the newly invented plastic, polyethylene. This material also proved to be unsatisfactory because it tended to age quickly, become brittle, and peel off the cables. Hawkins and his research partner, Vincent Lanza, developed a new type of plastic sheath for cables that solved all previous problems. It survived extremes of temperature and light and remained effective for periods of up to 70 years. It soon became the "go-to" material for telephone cable coverings.

Walter Lincoln Hawkins was born on March 21, 1911, in Washington, D.C. His father was an attorney for the U.S. Census Bureau, and his mother was a science teacher in the District of Columbia school system. He is said to have been intrigued by the construction and operation of machines at an early age and to have built his own radio receiver in order to listen to the broadcast of baseball games involving the Washington Senators. He attended the prestigious, all-Black Dunbar High School in Washington, where he was encouraged to continue his pursuit of science by one of his teachers, Dr. James Cowan.

Upon graduation from Dunbar, Hawkins matriculated at Rensselaer Polytechnic Institute, from where he earned his BS in chemical engineering in 1932. Normally, that accomplishment would not have been remembered as a particularly remarkable achievement. But at the time, college opportunities for African Americans were not particularly favorable, and a bachelor's degree represented perseverance as much as academic skill. Hawkins then continued his studies at Harvard University, from where he received his MS in chemistry in 1934, and at McGill University in Montréal in 1938. He then remained at McGill as instructor in chemistry from 1938 to 1941 and, from 1940 to 1942, also took part in a postdoctoral program in chemistry at Columbia University.

In 1942, Hawkins accepted a job offer at the Bell Telephone Laboratories (now AT&T Bell Laboratories), where he was to remain until his retirement in 1976. He was at the time, the first African American to join the technical staff at Bell. During his tenure at Bell, Hawkins continued to study the physical and chemical properties of plastics. His special interest eventually centered on the invention of plastic materials with longer life spans as one approach to the growing problem of plastic pollution that he saw around him. At Bell, Hawkins served as Head of Plastics Chemistry R&D, Supervisor of Applied Research, and eventually, head of plastic research at the company. After leaving Bell, Hawkins accepted an appointment as director of research at the Plastic Institute of America, where he remained until 1983. He also worked as consultant and expert witness on issues of plastic materials. He died in San Marcos, California, on August 20, 1992.

Throughout his career, Hawkins devoted his efforts to the advancement of Black and other disadvantaged youth in the educational system. In 1974, he was strongly involved in the development of the Bell Laboratories' Summer Research Program for Minorities and Women, and, in 1981, he became the first chairman of the American Chemical Society Subcommittee for the Education and Employment of the Disadvantaged (Project SEED). Among his numerous honors were the Burton C. Belden Award of the American Chemical Society, Percy L. Julian Award of the National Organization of Black Chemists, the International Award of the Society of Plastics Engineers, the Honor Scroll of the American Institute of Chemists, and the Achievement Award of the Los Angeles Council of Black Professional Engineers. He also received honorary doctorates from Montclair State College, Stevens Institute of Technology, Kean State College, and Howard University. In 1992, Hawkins was honored with the National Medal of Technology from President Bush at the White House.

## Stephanie Kwolek (1923–2014)

Stephanie Kwolek is one of only three women among more than 200 men in the Plastics Hall of Fame. She was recognized by this honor for her discovery of polyamide plastics of extraordinary strength. The most famous discovery resulting from this research was Kevlar, a fabric-like product that still has many applications today, including sports protection equipment and shoes, bicycle tires, police and military armor, string instruments, kitchenware, building construction materials, shoes, brakes, vehicle parts, and ropes and cables.

Stephanie Louise Kwolek was born on July 31, 1923, in New Kensington, Pennsylvania. Her parents were John and Nellie (Zajdel) Kwolek, Polish immigrants to the United States. John was a foundry worker who died when Stephanie was only 10 years old, and her mother was a seamstress. She attributed her interest in chemistry to her father's love for natural sciences and her mother's work with fabrics. After graduating from New Kensington High School, Kwolek matriculated at the Margaret Morrison Carnegie College of Carnegie Mellon University, where she hoped to major in pre-medicine. Over time, however, she became more interested in the study of chemistry and graduated from Margaret Morrison Carnegie in 1946 with a bachelor of science degree in that field.

Upon graduation, Kwolek accepted a position as research chemist at the DuPont company. She had originally considered the DuPont offer as a temporary job to make money for further studies in chemistry. And she was able to get the position because of the severe shortage of male chemists because of World War II. That "temporary" job at DuPont ended up lasting until 1986, when Kwolek retired from her post as research chemist at the company. By that time, she had received more than two dozen (some sources say 17) patents for her inventions and discoveries, most in connection with her work on Kevlar. After her retirement, she tutored several high school students in chemistry and wrote a variety of teaching tools and demonstrations in the field. One of those demonstrations is the so-called nylon rope trick, which is still used in many chemistry classrooms today.

During and following her academic career, Kwolek was honored with a number of awards, including the Chemical Pioneer Award of the American Institute of Chemists, Award for Creative Invention from the American Chemical Society (1995), National Medal of Technology (1996), IRI Achievement Award (1996), and the Perkin Medal from the American Chemical Society (1997). She also received honorary degrees from Worcester Polytechnic Institute (1981), Clarkson University (1997), and Carnegie Mellon University (2001). She was inducted into the National Inventors Hall of Fame in 1995 and the National Women's Hall of Fame in 2003. Great Britain's Royal Society of Chemistry established the Stephanie L Kwolek Award in 2014 "to recognize exceptional contributions to the area of materials chemistry from a scientist working outside the UK." Kwolek died in Wilmington, Delaware, on June 18, 2014, at the age of 90.

## Charles J. Moore (Dates Unavailable)

Moore has become famous because of a single very special event in his life: the discovery of the Great Pacific Garbage Patch in the Pacific Ocean in 1997. Moore grew up in Long Beach, California, home to his family for three generations. His father was an industrial chemistry by occupation and an avid sailor by off-work hobby. He often took his family to sea to destinations ranging from Guadalupe Island, Mexico, to points in Hawaii. Moore attended the University of San Diego (USD), where he majored in chemistry, while also studying mathematics, physics, and Spanish literature.

After leaving USD, he opened his own woodworking and finishing shop, a business he ran for 25 years with considerable success. In 1994, he left the business to create a new oceanographic entity, the Algalita Marine Research Foundation, in Long Beach. The mission of the foundation was to carry out research on the coastal waters of Southern California and work for the restoration of that natural environment. He soon became involved in a variety of oceanographic programs, first as a coordinator of the State Water Resources Control Board's Volunteer Water Monitoring Steering Committee and later as a member of the Southern California Water Research Projects' Bight '98 Steering Committee. In the latter position, he provided his own vessel as a platform for researchers conducting an assessment of the Southern California coastline from Point Concepcion to Ensenada. As part of that project, he also was instrumental in the development of a system for measuring microplastic particles that is still in use worldwide today.

In 1995, Moore built a new research vessel, the *Alguita*, which he first launched from Hobart, Tasmania. On its maiden voyage, the *Alguita* was damaged and had to return to Long Beach for repairs. When that work was completed, he decided to test the vessel by entering the Trans-Pac Yacht Race, from Los Angeles to Honolulu. After completing the race, he decided to take a "leisurely" route back to California, a trip that took him through a region known as the North Pacific Subtropical Gyre (GPGP). The North Pacific Subtropical Gyre is a region in the Pacific Ocean ranging from the western coast of the United States to the western coast of Central America, to the Philippine Islands and Japan. Its total size is just less than 8 million square miles. It was within this area that Moore, for the first time, saw a phenomenon now known as the Great Pacific Garbage Patch. It was an area, Moore later said, was notable not for "continuous scraps as it is now, but stray pieces here and there." "I could stand on deck for five minutes," he went on, "seeing nothing but the detritus of civilization in the remotest part of the great Pacific Ocean" ("Profile." n.d. http://www.captain-charles-moore.org/about).

Moore's experience in the GPGP transformed his life. He decided that he had to devote his work not just to the California coast, but to the much larger problem of plastic pollution in the ocean. The Alguita Foundation has now sponsored more than 10 trips into the North Pacific Gyre to learn more about the extent and composition of the GPGP. Moore himself has written two scholarly papers on the topic and has become a speaker in worldwide demand on the topic.

### Alexander Parkes (1813–1890)

The question is often raised as to who is the "founder" of the plastics industry. One name frequently mentioned is that of John Wesley Hyatt (1837–1920), an American inventor who first made the material known as celluloid. He invented that product in response to a shortage of natural ivory, the material most commonly used in the early nineteenth century for billiard balls. His invention involved combining a mixture combination of cellulose nitrate, camphor, and alcohol and then heating that mixture and pouring it into a mold. He and his brother Isaiah obtained a patent for the product in 1870 and later established the Celluloid Manufacturing Company to market their product.

But Hyatt was actually preceded by a few years by the work of British inventor Alexander Parkes. Parkes explored the possibility of creating a new synthetic rubber-like product by reacting natural cellulose (a major component of plants) with nitric acid. The major problem in his research was to discover a solvent in which the product of this reaction, cellulose nitrate (also known as nitrated cellulose, pyroxylin, or by other names), could be dissolved in order to produce a material with the desired physical properties. He eventually found a variety of chemicals that satisfied this requirement, and, in 1850, he received his patent for the material, often referred to as the first plastic. In his patent application, Parkesine referred to his product as a form of both colloidon and nitrated cellulose, but he eventually settled on Parkesine as the commercial name of his product. Parkes later established a company to market his product, but was not much of a success as a businessman, and his business failed. Parkesine remains popular today for a limited variety of uses, most commonly in the manufacture of jewelry and other decorative objects.

Alexander Parkes was born on December 29, 1813, in Birmingham, England, the fourth of eight children. His father was a locksmith, and young Alexander soon developed an interest in that profession and in the operation of mechanical devices in general. As a young man, he was apprenticed to Messenger and Sons, a brass foundry in Birmingham. He later joined the George and Henry Elkington silver-plating business (later, Elington and Mason). There he invented several of the 80 metallurgical products and methods for which he earned patents.

In 1866, Parkes established a company to sell his new product, Parkesine, but the venture lasted only a short time. Throughout his life, he was far more interested in inventing and exploring new ideas in materials than in commercialization of his discoveries. One biography has suggested that "he could never really be successful in business as he had too many irons in the fire and his mind was never still, always thinking of the next idea" (Richard Edmunds and Ellen Davies. 2019. "Men Who Made Burry Port." https://pembreyburryportheritage.co .uk/home/men-made-burry-port).

Parkes was also productive as a husband of two wives, the first of whom gave him 8 children and the second 12 children. After the death of his wife, he moved to Burry Port, Wales, where he continued his work in metallurgy and related fields. He then moved back to Birmingham to continue his works and, in 1866, to London, where he opened his ill-fated Parkesine Company. He died in West Dulwich, near London, on June 29, 1890.

Parkes was always much involved over the controversy as to who it was that actually invented the first plastic, celluloid— himself or Hyatt. At one point, he wrote, "I do wish the World to know who the inventor [of celluloid] really was, for it is a poor reward after all I have done to be denied the merit of the invention" (Sue Mossman. 2013. "Alexander Parkes—Materials Man and Polymath." Science Museum. https://blog.sciencemuseum .org.uk/alexander-parkes-materials-man-and-polymath/). That point of contention remains even today among historians of science.

## Boyan Slat (1994–)

One of the most extensive problems of solid waste management in the world today is the presence and persistence of plastics in the oceans. The most dramatic example of this problem is the Great Pacific Garbage Patch (GPGP), a region of the Pacific Ocean that may be at least as large as twice the size of France. Scholars have been thinking about possible ways of getting rid of the GPGP ever since its discovery in 1997. Most solutions are based on ways of attacking plastic pollution at its source, by reducing the amount of plastic produced around the world, developing mechanisms for degrading plastics before they are deposited into waterways, and educating the general public about the risks of ocean pollution by plastics.

None of these ideas help with ocean plastic pollution that already exists. It would be helpful if the rate of plastic disposal could be reduced in the future, but what about the vast quantities of plastic currently present in the ocean? Some ideas have been suggested for designing giant vacuum-like devices that could suck plastics out of the oceans or for finding and using microorganisms that can degrade plastics (few of which are currently known). In 2011, a Dutch teenage boy, Boyan Slat, came up with a plan for removing in situ plastic wastes from the oceans. He first hit upon the idea while taking diving lessons on the Greek island of Lesbos in 2010. During his classes, he noted that he actually saw more plastic waste in the oceans than he did fish. This struck him as an idea for a high school science project, which he completed the following year.

The project was based on a system of floating plastic tubes capable of collecting wastes and transferring them to a central storage and disposal site. The idea was a great success; it won the local science fair award, and Slat decided to continue his studies at the Delft Technical University (TUDelft) in aerospace engineering. His plans were sidetracked, however, when he was invited to given a TED talk about his invention. Although his Delft audience was relatively small, the video of the talk went viral, and Slat soon became something of a celebrity in the field of plastic waste management. He decided to withdraw from TUDelft and focus all his attention on the GPGP cleansing scheme. In 2013, he founded The Ocean Cleanup, an organization of which he is now CEO (chief executive officer). The organization got its start by way of a crowdfunding campaign that produced $2.2 million from 38,000 donors in 160 countries.

Slat was born in Delft, The Netherlands, on July 27, 1994. His father was an artist, and his mother worked in the tourism industry. His parents were divorced when he was still a baby, and he continues to live in Delft as of 2019. He has already begun to amass awards for his ideas for plastic cleanup. In 2014, he received the United Nations' Champion of the Earth award, the youngest recipient ever for that honor. Among his other honors include the Young Entrepreneur Award for 2015 from King Harald of Norway, a Thiel fellowship endowed by PayPal co-founder Peter Thiel, a European of the Year award from *Reader's Digest* magazine in 2017, and a European Entrepreneur of the Year award for 2018 by Euronews television news network.

## Surfrider Foundation

The Surfrider Foundation is a 501(c)(3) nonprofit organization that works to protect the world's oceans, beaches, and waves. It was founded in 1984 by a group of surfers led by Chris Blakely, Lance Carson, Glenn Hening, and Tom Pratte. The men surfed regularly in Malibu, California, and had become aware of the environmental hazards developing in that area. They decided to form an organization that would work to protect the Malibu beaches, as well as beaches around the world, from the dangers they faced. The name "Surfrider" does not come, as one might expect, from the shared recreational hobby, but from the name of a motel at which Hening had stayed as a boy.

One of the organization's first victories came in a court case against the U.S. Army Corps of Engineers, which was planning the construction of a breakwater at Imperial Beach, in southern California. Surfrider argued that the new structure would do harm to the beaches in the area, a contention with which the court agreed, and the breakwater project was canceled. The Imperial Beach action was soon followed by similar court cases at other California beaches, including Santa Maria Rivermouth in Santa Barbara county, South Cardiff State Beach in San Diego county, and Bolsa Chica State Park at Huntington Beach.

Surfrider has grown to a membership of more than 50,000 paid subscriptions, which are organized into more than 80 chapters worldwide. Most chapters are located in the United States, 19 chapters in California, 6 in Washington state, 7 on the Gulf Coast, 29 along the Eastern seaboard, and 8 in foreign countries, including Australia, Brazil, Canada, Japan, and New Zealand.

Surfrider carries out its goal through a group of initiatives and programs. Among the former are Clean Water, Beach Access, Ocean Protection, Coastal Preservation, and Plastic Pollution efforts. These initiatives are associated with and carried out through five targeted programs: Blue Water Task Force, Ocean Friendly Gardens, Rise above Plastic, Beach Cleanups, and Ocean Friendly Restaurants. The first of these programs is designed to collect and analyze information about water quality that can be provided to local communities, helping them to form better pollution prevention and treatment programs in their own setting. Ocean Friendly Gardens is a program that teaches local communities how to use plant life to improve water quality in their region. Activities involve the construction of aquatic gardens to filter out water pollutants that would otherwise empty into the oceans, restore lost and damaged soils, and restore wildlife refuges.

The Rise above Plastics program was created in response to the growing threat to water resources posed by plastic pollution. Local groups are provided with educational materials to be distributed and used in aiding individual communities to understand and act on the hazards posed by increasing levels of plastic in the oceans. Beach Cleanups are regular activities conducted on the East and West Coasts of the United States, as well as the coasts of Hawaii and Puerto Rico. The Ocean Friendly Restaurants program recognizes that oceanside restaurants can be major contributors to ocean pollution. Surfrider helps such businesses to identify their own activities that form the basis of this problem and aids them in developing more environmental-friendly and sustainable operating practices.

Much of the history of plastics, including debates over the topic in today's world, is revealed by a study of documents on the topic: laws, legislative bills, court cases, administrative rulings, and the like. This chapter contains excerpts from a dozen such items. They provide insights into the ways in which legislators, judges, and the common man and woman understand the benefits that plastics have to offer in our society, as well as the problems plastics present for the community and some ways of dealing with those problems. In addition, the chapter contains nine tables that provide quantitative information about the production, use, and disposal of plastics.

## Data

### Table 5.1. Global Production of Plastics, 1950–2015

Table 5.1 provides a summary of estimated values for the amount of plastic produced throughout the world from 1950 to 2015. The numbers given are only estimates, since reliable data on the topic are not available.

---

One important element in dealing with the problem of plastics pollution is recycling. However, the rate of plastics recycling in the United States remains very low. (Shutterstock.com)

Table 5.1    Global Production of Plastics, 1950–2015

| Year | Production (in tonnes) |
|------|----------------------|
| 1950 | 2 million |
| 1955 | 4 million |
| 1960 | 8 million |
| 1965 | 17 million |
| 1970 | 35 million |
| 1975 | 46 million |
| 1980 | 70 million |
| 1985 | 90 million |
| 1990 | 120 million |
| 1995 | 156 million |
| 2000 | 213 million |
| 2005 | 263 million |
| 2010 | 313 million |
| 2015 | 381 million |

Source: Ritchie, Hannah, and Max Roser. n.d. "Global Plastics Production." Our World in Data. https://ourworldindata.org/plastic-pollution#plastic-waste-per -person.

## Table 5.2. Disposal of Plastic Wastes, 1980–2015 (%)

Table 5.2 summarizes the fate of plastic wastes—in landfills, incineration, and recycling—worldwide between 1980 and 2015.

Table 5.2    Disposal of Plastic Wastes, 1980–2015 (%)

| Year | Disposal Method | | |
|------|-----------|-------------|----------|
| | Discarded | Incinerated | Recycled |
| 1980 | 100.00 | 0.00 | 0.00 |
| 1985 | 95.50 | 4.50 | 0.00 |
| 1990 | 90.00 | 8.00 | 2.00 |
| 1995 | 83.00 | 11.50 | 5.50 |
| 2000 | 76.00 | 15.00 | 9.00 |
| 2005 | 69.00 | 18.50 | 12.50 |
| 2010 | 62.00 | 22.00 | 16.00 |
| 2015 | 55.00 | 25.50 | 19.50 |

Source: Ritchie, Hannah, and Max Roser. n.d. "Global Plastic Waste by Disposal." Our World in Data. https://ourworldindata.org/plastic-pollution#plastic -waste-per-person.

## Table 5.3. Plastic Use by Sector (million tonnes) (2015)

Table 5.3 provides an overview of the uses of plastic materials in general in the calendar year 2015.

**Table 5.3   Plastic Use by Sector (million tonnes) (2015)**

| Sector | Amount |
| --- | --- |
| Additives | 25 |
| Building and construction | 65 |
| Consumer and institutional products | 42 |
| Electrical/electronic | 18 |
| High-density polyethylene | 52 |
| Industrial machinery | 3 |
| Low-density polyethylene | 64 |
| Other polymer type | 16 |
| Other sector | 59 |
| Polyethylene terephthalate | 33 |
| Polypropylene | 68 |
| Polyphthalamide fibers | 59 |
| Polystyrene | 25 |
| Polyurethane | 27 |
| Polyvinylchloride | 38 |
| Packaging | 1,460 |
| Textiles | 47 |
| Transportation | 27 |

*Source*: Ritchie, Hannah, and Max Roser. n.d. "Global Plastic Waste by Disposal." Our World in Data. https://ourworldindata.org/plastic-pollution#plastic-waste-per-person.

## Table 5.4. Per Capita Plastic Waste for Selected Countries, 2010 (kilograms per person per day)

The amount of plastic products used worldwide varies substantially from country to country. Table 5.4 provides information on this item for selected nations.

Table 5.4   Per Capita Plastic Waste for Selected Countries, 2010 (kilograms per person per day)

| Country | Waste |
| --- | --- |
| Australia | 0.112 |
| Barbados | 0.57 |
| Belgium | 0.08 |
| British Virgin Islands | 0.252 |
| Canada | 0.093 |
| Chile | 0.119 |
| China | 0.121 |
| Costa Rica | 0.258 |
| Cuba | 0.089 |
| Denmark | 0.047 |
| Egypt | 0.178 |
| Finland | 0.234 |
| France | 0.192 |
| Germany | 0.485 |
| Greenland | 0.252 |
| Guyana | 0.586 |
| Haiti | 0.09 |
| Hong Kong | 0.398 |
| India | 0.01 |
| Israel | 0.297 |
| Italy | 0.134 |
| Japan | 0.171 |
| Kenya | 0.027 |
| Kuwait | 0.686 |
| Libya | 0.144 |
| Madagascar | 0.016 |
| Mexico | 0.087 |
| Mozambique | 0.015 |
| Netherlands | 0.424 |
| Nigeria | 0.103 |
| North Korea | 0.054 |
| Pakistan | 0.103 |
| Palestine | 0.063 |

(continued)

Table 5.4    (*continued*)

| Country | Waste |
| --- | --- |
| Philippines | 0.075 |
| Puerto Rico | 0.252 |
| Russia | 0.112 |
| Saint Kitts | 0.654 |
| Saudi Arabia | 0.156 |
| Singapore | 0.194 |
| Somalia | 0.054 |
| South Korea | 0.112 |
| Spain | 0.277 |
| Sweden | 0.048 |
| Thailand | 0.144 |
| Turkey | 0.212 |
| Ukraine | 0.103 |
| United Kingdom | 0.215 |
| United States | 0.335 |
| Vietnam | 0.103 |

*Source*: Ritchie, Hannah, and Max Roser. n.d. "Global Plastic Waste by Disposal." Our World in Data. https://ourworldindata.org/plastic-pollution#plastic-waste-per-person. This source also has an interesting graph showing Per Capita Plastic Waste versus Gross Domestic Product per Person.

## Table 5.5. Plastic Production and Disposal in the United States, Selected Years (thousands of tons)

Methods by which plastics are disposed of in the United States have varied substantially from 1960 to the present day. This table summarizes those trends.

Table 5.5  Plastic Production and Disposal in the United States, Selected Years (thousands of tons)

| Year | Production | Recycled | | Incineration with Energy Recovery | | Landfill | |
| --- | --- | --- | --- | --- | --- | --- | --- |
| | | Weight | Percent-age (%)* | Weight | Percent-age (%)* | | Percent-age (%)* |
| 1960 | 390 | n/a | — | n/a | — | 390 | 100.00 |
| 1970 | 2,900 | n/a | — | n/a | — | 2,900 | 100.00 |
| 1980 | 6,830 | 20 | 0.30 | 140 | 2.04 | 6,670 | 97.66 |
| 1990 | 17,130 | 370 | 2.15 | 2,980 | 17.40 | 13,780 | 80.44 |
| 2000 | 25,550 | 1,480 | 5.80 | 4,120 | 16.12 | 19,950 | 78.08 |
| 2005 | 29,380 | 1,780 | 6.06 | 4,330 | 14.74 | 23,270 | 79.20 |
| 2010 | 31,400 | 2,500 | 7.96 | 4,530 | 14.43 | 24,370 | 77.61 |
| 2015 | 34,480 | 3,120 | 9.05 | 5,330 | 15.46 | 26,030 | 75.49 |
| 2016 | 34,870 | 3,240 | 9.30 | 5,340 | 15.31 | 26,290 | 75.39 |
| 2017 | 35,370 | 2,960 | 8.37 | 5,590 | 15.80 | 26,820 | 75.83 |

n/a = not available
*Percentage = Percentage of Disposal Method (calculated by author)
Source: "National Overview: Facts and Figures on Materials, Wastes and Recycling." 2019. Environmental Protection Agency. https://www.epa.gov/facts -and-figures-about-materials-waste-and-recycling/national-overview-facts-and -figures-materials.

## Table 5.6. Microplastic Concentration at Various Points in the United States

Table 5.6 provides some basic data on the types of microplastics found at a group of U.S. ports, along with the hypothesized source of those materials.

Table 5.6  Microplastic Concentration at Various Points in the United States*

| Location | Urban | Agricultural | Other | Fiber & Lines | Films | Foams | Fragments | Beads & Pellets |
|---|---|---|---|---|---|---|---|---|
| St. Louis, MN | 2.9 | 2.9 | 94.3 | 0.3 | 0.0 | 0.0 | 0.0 | 0.0 |
| Nemadji, WI | 3.0 | 9.9 | 87.0 | 0.7 | 0.0 | 0.0 | 0.0 | 0.0 |
| Genessee, NY | 6.6 | 45.3 | 48.1 | 1.3 | 0.0 | 0.1 | 0.1 | 0.0 |
| Manitowoc, WI | 7.0 | 69.9 | 23.1 | 0.2 | 0.0 | 0.0 | 0.0 | 0.0 |
| Tonawanda, NY | 7.3 | 47.9 | 44.8 | 1.2 | 0.0 | 0.0 | 0.0 | 0.0 |
| Sandusky, OH | 8.3 | 80.5 | 11.2 | 0.9 | 0.0 | 0.1 | 0.1 | 0.0 |
| Fox, WI | 8.4 | 41.4 | 50.2 | 0.9 | 0.1 | 0.1 | 0.0 | 0.0 |
| Sheboygan, WI | 8.6 | 64.2 | 27.1 | 0.3 | 0.0 | 0.0 | 0.0 | 0.0 |
| Grand, OH | 8.9 | 34.3 | 56.8 | 0.5 | 0.0 | 0.0 | 0.2 | 0.0 |
| Huron, OH | 9.5 | 74.5 | 15.9 | 0.4 | 0.0 | 0.0 | 0.0 | 0.0 |
| Portage, OH | 9.8 | 84.2 | 6.0 | 0.6 | 0.0 | 0.0 | 0.0 | 0.0 |
| Maumee, OH | 10.7 | 78.7 | 10.5 | 0.9 | 0.0 | 0.0 | 0.0 | 0.0 |
| Paw Paw, MI | 11.0 | 48.1 | 40.0 | 2.0 | 0.0 | 0.0 | 0.1 | 0.0 |
| Raisin, MI | 11.1 | 67.3 | 21.6 | 0.5 | 0.0 | 0.0 | 0.1 | 0.0 |
| Ashtabula, OH | 11.8 | 35.3 | 52.0 | 4.5 | 0.1 | 0.0 | 0.3 | 0.0 |
| Saginaw, MI | 12.2 | 44.5 | 43.4 | 1.6 | 0.0 | 0.0 | 0.2 | 0.0 |
| Black, OH | 12.6 | 56.2 | 31.3 | 0.3 | 0.0 | 0.0 | 0.1 | 0.0 |
| Kalamazoo, MI | 13.6 | 48.7 | 37.7 | 0.6 | 0.0 | 0.0 | 0.1 | 0.0 |
| St. Joseph, MI | 13.8 | 60.4 | 25.8 | 0.2 | 0.0 | 0.1 | 1.0 | 0.0 |
| Grand, MI | 14.8 | 54.0 | 31.2 | 1.3 | 0.0 | 0.0 | 0.1 | 0.0 |
| Buffalo, NY | 16.3 | 34.1 | 49.5 | 1.6 | 0.2 | 1.2 | 0.5 | 1.0 |
| Huron, MI | 28.9 | 25.2 | 45.9 | 1.2 | 0.2 | 0.2 | 1.8 | 0.0 |

(continued)

**Table 5.6** (*continued*)

| Location | Urban | Agricultural | Other | Fiber & Lines | Films | Foams | Fragments | Beads & Pellets |
|---|---|---|---|---|---|---|---|---|
| Milwaukee, WI | 29.9 | 43.2 | 26.9 | 0.4 | 0.0 | 0.9 | 1.3 | 0.1 |
| Burns, IN | 38.0 | 27.4 | 34.7 | 0.0 | 0.0 | 0.1 | 0.0 | 0.0 |
| Cuyahoga, OH | 40.0 | 17.5 | 42.5 | 1.1 | 0.0 | 0.1 | 0.1 | 0.0 |
| Rocky, OH | 40.6 | 22.2 | 37.2 | 0.9 | 0.1 | 0.0 | 0.1 | 0.0 |
| Clinton, MI | 52.8 | 19.5 | 27.7 | 1.7 | 0.4 | 0.3 | 2.2 | 0.0 |
| Indiana City, IN | 84.1 | 1.0 | 14.9 | 0.5 | 0.0 | 0.1 | 2.0 | 0.1 |
| Rouge, MI | 92.0 | 0.1 | 7.9 | 2.4 | 0.1 | 0.6 | 0.8 | 0.0 |

*Land use: percentage of basin
Microplastics: particles per 100 gallons of water
*Source:* "Microplastics in Our Nation's Waterways." 2016. U.S. Geological Survey. https://owi.usgs.gov/vizlab/microplastics/.

## Table 5.7. Fate of All Plastics Produced Worldwide, 1980–2015

The ways in which plastics are disposed of have varied in the period from 1980 to the present day. This table provides a summary of those trends.

Table 5.7    Fate of All Plastics Produced Worldwide, 1980–2015

| Year | Recycled | Incinerated | Discarded |
|------|----------|-------------|-----------|
| 2015 | 0.192 | 0.247 | 0.561 |
| 2014 | 0.185 | 0.240 | 0.575 |
| 2013 | 0.178 | 0.220 | 0.602 |
| 2012 | 0.171 | 0.201 | 0.628 |
| 2011 | 0.164 | 0.148 | 0.688 |
| 2010 | 0.157 | 0.145 | 0.698 |
| 2009 | 0.150 | 0.141 | 0.709 |
| 2008 | 0.143 | 0.137 | 0.720 |
| 2007 | 0.136 | 0.133 | 0.731 |
| 2006 | 0.129 | 0.130 | 0.741 |
| 2005 | 0.122 | 0.126 | 0.752 |
| 2004 | 0.115 | 0.122 | 0.763 |
| 2003 | 0.108 | 0.118 | 0.774 |
| 2002 | 0.101 | 0.114 | 0.785 |
| 2001 | 0.094 | 0.110 | 0.796 |
| 2000 | 0.087 | 0.107 | 0.806 |
| 1999 | 0.080 | 0.103 | 0.817 |
| 1998 | 0.073 | 0.099 | 0.828 |
| 1997 | 0.066 | 0.095 | 0.839 |
| 1996 | 0.059 | 0.091 | 0.850 |
| 1995 | 0.052 | 0.087 | 0.860 |
| 1994 | 0.045 | 0.080 | 0.874 |
| 1993 | 0.038 | 0.074 | 0.888 |
| 1992 | 0.031 | 0.067 | 0.902 |
| 1991 | 0.024 | 0.060 | 0.916 |
| 1990 | 0.017 | 0.053 | 0.930 |
| 1989 | 0.014 | 0.047 | 0.939 |
| 1988 | 0.013 | 0.040 | 0.947 |
| 1987 | 0.011 | 0.034 | 0.955 |
| 1986 | 0.009 | 0.028 | 0.962 |
| 1985 | 0.008 | 0.022 | 0.970 |

*(continued)*

Table 5.7    (*continued*)

| Year | Recycled | Incinerated | Discarded |
|------|----------|-------------|-----------|
| 1984 | 0.006 | 0.016 | 0.978 |
| 1983 | 0.005 | 0.012 | 0.983 |
| 1982 | 0.003 | 0.008 | 0.989 |
| 1981 | 0.002 | 0.004 | 0.994 |
| 1980 | 0.000 | 0.000 | 1.000 |

*Source*: Geyer, Roland, Jenna R. Jambeck, and Kara Lavendar Laws. 2017. "Production, Use, and Fate of All Plastics Ever Made." Science Advances. 3: e1700782, Table S7. https://advances.sciencemag.org/content/3/7/e1700782. This table also includes projects to the year 2015. Used with kind permission of the authors.

## Table 5.8. Global Production of Various Types of Plastics for 2015 (millions of tons)

Table 5.8 lists the most common types of plastics produced in the world today, along with the estimated amount for each type of plastic.

Table 5.8    Global Production of Various Types of Plastics for 2015 (millions of tons)

| Plastic | Production |
|---------|-----------|
| Polypropylene (PP) | 68 |
| Low-density polyethylene (LDPE) | 64 |
| Polyester, polyamide, and acrylic fibers (PP&A) | 59 |
| High-density polyethylene (HDPE) | 52 |
| Polyvinylchloride (PVC) | 38 |
| Polyethylene terephthalate (PET) | 33 |
| Polyurethane (PUR) | 27 |
| Polystyrene (PS) | 25 |
| Plastic additives | 25 |
| Other plastics | 16 |

*Source*: Beckman, Eric. 2018. "The World's Plastic Problem in Numbers." World Economic Forum. https://www.weforum.org/agenda/2018/08/the-world-of-plastics-in-numbers.

## Table 5.9 Forms of Polyethylene

The most popular type of plastic, polyethylene, is made in a variety of forms with different physical properties and different acronyms. This table provides basic information on those topics.

**Table 5.9   Forms of Polyethylene**

| Name | Abbreviation | Density (g/cm$^3$) |
| --- | --- | --- |
| Ultra-high-molecular-weight polyethylene | UHMWPE | 0.930–0.935 |
| Ultra-low-molecular-weight polyethylene | ULMWPE; PE-WAX | 0.92–0.96[1] |
| High-molecular-weight polyethylene | HMWPE | 0.966–1.099[2] |
| High-density polyethylene | HDPE | ~0.941 |
| High-density cross-linked polyethylene | HDXLPE | 1.35–2.20[3] |
| Cross-linked polyethylene | PEX; XLPE | ~0.952[4] |
| Medium-density polyethylene | MDPE | 0.926–0.940 |
| Linear low-density polyethylene | LLDPE | 0.915–0.925 |
| Low-density polyethylene | LDPE | 0.910–0.940 |
| Very-low-density polyethylene | VLDPE | 0.880–0.915 |
| Chlorinated polyethylene | CPE | 0.93–0.96[5] |

*Sources*: "Plastics Info: Plastics Classification." 2019. Plastixportal. https://www
.plastixportal.co.za/plastic_materials_classifications_and_applications.html.   (All
non-superscripted sources.)

[1] "Polyethylene Waxes, Technical Data." 2020. Deurex. https://www.deurex.com
/waxes/polyethylenewaxes/technical-data/.

[2] "LUBMER High Molecular Weight Polyethylene." 2020. Mitsui Chemicals
America, Inc. https://www.mitsuichemicals.com/lubmer_prop.htm.

[3] "Poly Processing HDXLPE Vertical Bulk Storage Tanks." 2020. Aetna Plastics
Corp. https://www.aetnaplastics.com/products/d/PolyProcessingHDXLPE.

[4] Freitas, Rodrigo Sousa, and Baltus Cornelius Bonse. 2019. "Cross-linked
Polyethylene (XLPE) as Filler in High-density Polyethylene: Effect of Content and
Particle Size." AIP Conference Proceedings. https://aip.scitation.org/doi/pdf/10
.1063/1.5084810.

[5] "Chlorinated Polyethylene (CPE) Characteristics and Application." 2017. ViA
Chemical.   http://viachemical.com/chlorinated-polyethylene-cpe-characteristics
-and-application/.

## Documents

### Marine Plastic Pollution Research and Control Act (1987)

*For most nations of the world, plastic pollution is a relatively recent phenomenon. While solid waste management in general has long been an issue of significant proportions to almost all civilized cultures, the widespread use of plastics and their ultimate fate has not been a topic of concern until about the 1980s. It should be no surprise, then, that legislation at the state and federal level in the United States has been almost nonexistent. An example of one of the few legislative acts to mention plastics pollution even in recent history is the so-called Marine Plastic Pollution Research and Control Act of 1987 (MPPRCA). That act was adopted as part of a larger bill, the United States Japan Fishery Agreement Approval Act of 1987, and as an amendment to the earlier Act to Prevent Pollution from Ships of 1980. MPPRCA was a timid piece of legislation that focused primarily on the authorization of research on the problem of plastic pollution and possible methods for dealing with that problem. The following section is an excerpt of that act.*

SEC. 2201. COMPLIANCE REPORTS.

(a) IN GENERAL.—Within 1 year after the effective date of this section, and biennially thereafter for a period of 6 years, the Secretary of the department in which the Coast Guard is operating, in consultation with the Secretary of Agriculture and the Secretary of Commerce, shall report to the Congress regarding compliance with Annex V to the International Convention for the Prevention of Pollution from Ships, 1973, in United States waters.

(b) REPORT ON INABILITY TO COMPLY.—Within 3 years after the Contracts effective date of this section, the head of each Federal agency that operates or contracts for the operation of any ship referred to in section 3(bXlXA) of the Act to Prevent Pollution from Ships that may not be able to comply with the requirements of that section shall report to the Congress describing-

. . .

[The text then lists the conditions that must be reported to the Congress.]

. . .

### SEC. 2202. EPA STUDY OF METHODS TO REDUCE PLASTIC POLLUTION.

(a) IN GENERAL.—The Administrator of the Environmental Protection Agency, in consultation with the Secretary of Commerce, shall commence a study of the adverse effects of the improper disposal of plastic articles on the environment and on waste disposal, and the various methods to reduce or eliminate such adverse effects.

(b) SCOPE OF STUDY.—A study under this section shall include the following:

[Some contents of the study are listed, such as a list of improper disposal practices, a description of legislative relief needed to deal with the problem, a review of the role of plastics in the overall issue of solid waste management, and an analysis of using degradable plastics to reduce plastics pollution in the oceans.

. . .

### SEC. 2203. EFFECTS OF PLASTIC MATERIALS ON THE MARINE ENVIRONMENT.

Not later than September 30, 1988, the Secretary of Commerce shall submit to the Congress a report on the effects of plastic materials on the marine environment. The report shall—

[Some elements of the report are listed next, such as harmful effects of plastics on marine organisms, effects of plastics on specific organisms, types of plastics likely to pose the greatest threat to the marine environment, recommendations for ways of taxing harmful plastic products.

. . .

### SEC. 2204. PLASTIC POLLUTION PUBLIC EDUCATION PROGRAM.

The two types of programs authorized are "Outreach Programs" and "Citizen Pollution Patrols." The former are to focus on

. . .

(A) the harmful effects of plastic pollution;

(B) the need to reduce such pollution;

(C) the need to recycle plastic materials; and

(D) the need to reduce the quantity of plastic debris in the marine environment.

[Authorized activities to achieve these goals include]

. . .

(A) workshops with interested groups;

(B) public service announcements;

(C) distribution of leaflets and posters; and

(D) any other means appropriate to educating the public.

**Source:** Public Law 100-220. 1987.

### Policy Memorandum 11-03 ("Water Bottle Ban") (2011); Discontinuation of Policy (2017)

*Plastic water bottles are one of the major forms of plastics commonly disposed of. As one of its steps in trying to deal with the problem of plastics pollution in the United States, the U.S. National Park Service issued a memorandum in 2011 designed to reduce or eliminate the use of plastic bottles in the nation's national parks. That memorandum was discontinued under the administration of President Donald Trump in 2017. The original document and its reversal are cited here.*

The heart of our sustainability program is our comprehensive Green Parks Plan (GPP). The GPP will address water and energy use, green house gas emissions, reduction of waste streams, construction practices, as well as other issues, and sets goals that we will reach by 2016. The GPP is undergoing final review and will be released early next year. In light of recent interest in one element of the GPP, we are issuing the attached specific policy on the reduction/recycling of disposable plastic water bottles. It allows superintendents to halt the sale of these bottles if they (1) complete a rigorous impact analysis including an assessment of the effects on visitor health and safety, (2) submit a request in writing to their regional director, and (3) receive the approval of their regional director.

. . .

Reduction: Parks are strongly encouraged to reduce the sale of disposable plastic water bottles through visitor education on the environmental impact of purchasing decisions and the availability of reasonably priced reusable bottles which can be filled at water fountains or bottle refill stations.

Elimination: Where appropriate, superintendents may request approval from their regional director to eliminate the sale of water in disposable plastic bottles by analyzing and addressing the following factors-in writing:

- Amount of waste eliminated and pros/cons to overall park operations
- Infrastructure costs and funding source(s) for filling stations
- Contractual implications on concessioners, including considerations of new leaseholder surrender interest or possessory interest
- Operational costs of filling stations including utilities and regular public health testing
- Cost and availability of BPA free reusable containers
- Effect on concessioner and cooperating association sales revenue
- Availability of water within concession food service operations
- Visitor education in the park and online so that visitors may come prepared with their own water bottles
- Results of consultation with NPS Public Health Office
- A sign plan so that visitors can easily find filling stations
- Safety considerations for visitors who may resort to not carrying enough water or drinking from surface water sources with potential exposure to disease

- A system for annual evaluation of the program, including public response, visitor satisfaction, buying behavior, public safety, and plastic collection rates
- Results of consultation with concessioners and cooperating associations
- Timeline of phase in period

Education: Parks must develop a proactive visitor education strategy that addresses visitor expectations and explains the rationale for whatever plastic bottle reduction, recycling, or elimination effort is implemented.

**Source:** "Recycling and Reduction of Disposable Plastic Bottles in Parks." 2011. National Park Service. https://www.nps.gov /policy/PolMemos/PM_11-03.pdf.


### *Discontinuation of Policy*

In its commitment to providing a safe and world class visitor experience, the National Park Service is discontinuing Policy Memorandum 11 03, commonly referred to as the "Water Bottle Ban."

The 2011 policy, which encouraged national parks to eliminate the sale of disposable water bottles, has been rescinded to expand hydration options for recreationalists, hikers, and other visitors to national parks. The ban removed the healthiest beverage choice at a variety of parks while still allowing sales of bottled sweetened drinks. The change in policy comes after a review of the policy's aims and impact in close consultation with Department of the Interior leadership.

"While we will continue to encourage the use of free water bottle filling stations as appropriate, ultimately it should be up to our visitors to decide how best to keep themselves and their families hydrated during a visit to a national park, particularly during hot summer visitation periods," said Acting National Park Service Director Michael T. Reynolds.

Currently only 23 of the 417 National Park Service sites have implemented the policy. The revocation of the memorandum, which was put in place on December 14, 2011, is effective immediately. Parks will continue to promote the recycling of disposable plastic water bottles and many parks have already worked with partners to provide free potable water in bottle filling stations located at visitor centers and near trailheads.

**Source:** "National Park Service Ends Effort to Eliminate Sale of Disposable Water Bottles." 2017. National Park Service. https://www.nps.gov/orgs/1207/08-16-2017-water-bottles.htm.

### Microbead-Free Waters Act of 2015

*Among the very few federal laws that have been passed for the control of pollution by plastics and microplastics is the Microbead Free Waters Act, adopted in 2015. The complete act is reprinted here.*

SECTION 1. SHORT TITLE.
This Act may be cited as the "Microbead Free Waters Act of 2015".
SEC. 2. PROHIBITION AGAINST SALE OR DISTRIBUTION OF RINSE OFF
COSMETICS CONTAINING PLASTIC MICROBEADS.
(a) IN GENERAL.—Section 301 of the Federal Food, Drug, and Cosmetic Act (21 U.S.C. 331) is amended by adding at the end the following:
"(ddd)(1) The manufacture or the introduction or delivery for introduction into interstate commerce of a rinse off cosmetic that contains intentionally added plastic microbeads.
"(2) In this paragraph—
"(A) the term 'plastic microbead' means any solid plastic particle that is less than five millimeters in size and is intended to be used to exfoliate or cleanse the human body or any part thereof; and
"(B) the term 'rinse off cosmetic' includes toothpaste.".
(b) APPLICABILITY.

(1) IN GENERAL.—The amendment made by subsection (a) applies-

(A) with respect to manufacturing, beginning on July 1, 2017, and with respect to introduction or delivery for introduction into interstate commerce, beginning on July 1, 2018; and

(B) notwithstanding subparagraph (A), in the case of a rinse off cosmetic that is a nonprescription drug, with respect to manufacturing, beginning on July 1, 2018, and with respect to the introduction or delivery for introduction into interstate commerce, beginning on July 1, 2019.

(2) NONPRESCRIPTION DRUG.—For purposes of this subsection, the term "nonprescription drug" means a drug not subject to section 503(b)(1) of the Federal Food, Drug, and Cosmetic Act (21 U.S.C. 353(b)(1)).

(c) PREEMPTION OF STATE LAWS.—No State or political subdivision of a State may directly or indirectly establish under any authority or continue in effect restrictions with respect to the manufacture or introduction or delivery for introduction into interstate commerce of rinse off cosmetics containing plastic microbeads (as defined in section 301(ddd) of the Federal Food, Drug, and Cosmetic Act, as added by subsection (a)) that are not identical to the restrictions under such section 301(ddd) that have begun to apply under subsection (b).

(d) RULE OF CONSTRUCTION.—Nothing in this Act (or the amendments made by this Act) shall be construed to apply with respect to drugs that are not also cosmetics (as such terms are defined in section 201 of the Federal Food, Drug, and Cosmetic Act (21 U.S.C. 321)).

**Source:** Public Law 114-114. 2015.

## Effects of Microplastics on Aquatic Organisms (2016)

*In 2016, the U.S. Environmental Protection Agency commissioned a study to determine the toxicological effects of microplastics on various types of marine organisms. The conclusions reached by researchers in this study are as follows (citations to studies used in the study are omitted here).*

Plastic particles are ubiquitous in the aquatic environment and are routinely found along beaches, in sediment, within the water column, and at the water surface. Plastic debris can be found in freshwater and marine environments ranging from coastlines near densely populated areas to the remote open ocean and along remote island shorelines. The very characteristics that make plastics useful (e.g., durability and longevity) allow plastics to persist in the aquatic environment. Most plastics (by particle number) encountered in the aquatic environment are microplastics, which can be roughly the size of a grain of sand or a planktonic organism. Methods to extract, isolate and identify these microplastics exist, but standardization to enumerate and identify these very small particles need to be standardized. Numerous research studies demonstrate that plastics are ingested (either directly or via prey) by aquatic invertebrates, fish, seabirds, sea turtles, and marine mammals.

Plastics in aquatic systems contain chemicals originating from the plastic material, chemicals added during the manufacturing process, as well as organic chemicals, metals, and other contaminants absorbed from the water column. Given that many of these chemicals have been found to have harmful effects once in the aquatic environment, the potential toxicological impacts of these chemicals associated with plastic once ingested by aquatic organisms and aquatic dependent wildlife is an area of concern. However, aquatic organisms and seabirds face a multitude of environmental stressors and attributing toxicological impacts directly to the ingestion of plastics and associated contaminants is challenging because organisms are exposed to metals, and organic chemicals including PBT contaminants from wastewater discharges, atmospheric deposition, and other sources in addition to plastics. The extent to which plastics are a relative source of metals and other chemicals to aquatic organisms and aquatic dependent wildlife is a recent area of study.

There is evidence that aquatic organisms and aquatic dependent wildlife accumulate chemicals from ingested plastics. Field studies have observed correlations of plastic densities or chemical concentrations in plastic with chemical concentrations in organisms and laboratory experiments document transfer of chemicals from plastic to organisms when there is a concentration gradient favoring transfer (i.e., high concentrations of chemicals on plastics and/or unexposed experimental organisms). Limited modeling approaches have been used to attempt to mimic environmentally realistic scenarios, and the models generally show a small to negligible contribution of plastic to the bioaccumulation of associated chemicals to aquatic organisms and seabirds, relative to other sources. Some recent reviews suggest that, the bioaccumulation of chemicals associated with plastics is most likely overwhelmed by uptake through other pathways; however, this does not imply that plastics do not have negative effects on aquatic organisms. A limited number of toxicological studies have been performed, mainly in the laboratory, investigating the effects of chemicals associated with plastics. Similar to the laboratory bioaccumulation studies, many of the toxicological studies were conducted under environmentally unrealistic situations that favored accumulation of the chemicals from the plastic. While negative sublethal effects were observed in treatments with chemically contaminated plastics and effects were often greater than treatments with the plastic alone, adverse effects were also demonstrated in organisms exposed to the plastic alone.

**Source:** Beaman, Joe, et al. 2016. "State of the Science White Paper: A Summary of Literature on the Chemical Toxicity of Plastics Pollution to Aquatic Life and Aquatic Dependent Wildlife." U.S. Environmental Protection Agency, 36–37. https://www.epa.gov/wqc/white-paper-summary-literature -chemical-toxicity-plastics-pollution-aquatic-life-and-aquatic.

### No More Plastics in the Oceans (2017)

*Adelphi is an independent think tank and public policy consulting agency dealing with problems relating to climate, environment, and development. They offer a variety of services in the areas of sustainable development and issues relating to global challenges for governmental agencies, international organizations, businesses, and nonprofit organizations. The Ecologic Institute is a think tank working on problems of the environment and public policy. In 2018, the two groups released a Discussion Paper on the problem of plastics pollution in the world's oceans. The following selection is a summary of the groups' findings on the nature of the problem and possible solutions for it. (footnotes omitted)*

Although there are at least three globally binding agreements dealing with sea based sources of marine litter (UNCLOS, MARPOL, and the London Convention), two multilateral environmental conventions addressing trade in hazardous waste and persistent organic pollutants (the Basel and Stockholm Convention), 18 regional seas programmes, some of which contain legally binding stipulations against marine litter, and a range of partnerships and other commitments, including the Global Programme of Action (GPA) and its Global Partnership on Marine Litter (GPML), glaringly obvious gaps remain in the governance structure of marine plastic pollution. The major gaps identified can be described as follows:

1. There is no agreement effectively preventing and minimizing marine plastic pollution, particularly from land based sources. Rather, there is a large number of frameworks which address various aspects of the problem. However, many of these have compliance and implementation problems, lack quantified goals, and most sources of plastic pollution in the ocean remain unregulated.

2. There is a lack of resources and technical cooperation, particularly on efforts to improve waste collection systems, enabling and enhancing qualitative recycling, fostering national and local regulatory systems, monitoring compliance

with existing regulations at the national level and on supporting the establishment of additional efforts.

3. There is a lack of coordination among the various frameworks, instruments and platforms dealing with plastic pollution. While stronger coordination may contribute to narrowing some of the gaps, there is a need for significantly enhanced cross sectoral coordination, for substantially boosted multi stakeholder driven collaborative efforts and for much closer intergovernmental cooperation.

4. There is no institutionalised effort for assessing the state of plastic pollution, and a lack of standards for monitoring the release of plastic waste or for the current state of plastic waste in the environment, including oceans. As a consequence, there are considerable uncertainties about the amounts, sources and transmission pathways of marine plastic pollution.

A new legally binding international agreement would be essential to closing these gaps and to effectively addressing marine plastic pollution. A convention on the elimination of marine plastic pollution should contain the following essential elements:

1. A clear goal to stop further marine plastic pollution by prohibiting the discharge of plastic waste into the ocean from land and sea based sources. This would build on UNEA Resolution 3/7 outlining "the importance of long term elimination of the discharge of litter and microplastics into the ocean".

2. Binding national plastic pollution reduction targets which address all sources and outline clearly the responsibilities of governments. The reduction targets should be connected to national plans for action, which should also address the responsibilities of business.

3. A technical cooperation and financing mechanism, providing the means of implementation and technical assistance for adopting a range of tools on the regional, national, and local level, for instance supporting sustainable financing of waste management systems, by schemes based on extended producer responsibility, for example.

4. A follow up and review mechanism, as well as an enhanced science policy interface, are needed for tracking progress towards objectives and targets and for continued observation of environmental conditions. The treaty should contain measures to assess conditions in the marine environment in order to monitor implementation of the treaty.

5. A central forum for coordination and for establishing partnerships among governments and other stakeholders. The convention's decision making body and other platforms would also bring actors from the various existing platforms and frameworks together to develop programmes and make use of synergies to address the full life cycle of plastics.

**Source:** Simon, Nils, and Maro Luisa Schulte. 2017. "Strengthening Plastic Governance: Towards a New Global Convention." Berlin, Germany: Adelphi. Used by kind permission of Dr. Simon.

### *City of Laredo, Texas, Petitioner, v. Laredo Merchants Association* (NO. 16-0748) (2018)

*In 2014, the city council of Laredo, Texas, passed an ordinance designed to reduce the use of one-time-use paper and plastic bags. The council offered three reasons for its actions:*

(a) To promote the beautification of the city through prevention of litter generated from discarded checkout bags.

(b) To reduce costs associated with floatable trash controls and the maintenance of the municipal separate stormwater sewer system.

(c) To protect life and property from flooding that is a consequence of improper stormwater drainage attributed in part to obstruction by litter from checkout bags (as cited in the court decision, NO. 16-0748, 5).

*Before the ordinance could take effect, the Laredo Merchants Association filed suit against the city, claiming that the ordinance was a violation of the Texas Constitution. The section to which the suit referred is Section 361.0961, which says that*

"A local government or other political subdivision may not adopt an ordinance, rule, or regulation to:
(1) prohibit or restrict, for solid waste management purposes, the sale or use of a container or package in a manner not authorized by state law;" Health and Safety Code 361.0961. https://statutes.capitol.texas.gov/StatutesByDate.aspx?code=HS&level=SE&value=361.0961&date=2/25/2015.

*The question before the court is whether the state constitution had priority in this matter and whether the Laredo ordinance was unconstitutional. The court's line of reasoning was as follows. (Citations are omitted.)*

The Act thus describes a state interest in "controlling the management of solid waste" that is plenary. The Act's preemption of local control is narrow and specific, applying to ordinances that "prohibit or restrict, [1] for solid waste management purposes, [2] the sale or use of a container or package [3] in a manner not authorized by state law". The City argues that its Ordinance does not meet any of these elements. We address each in turn.

. . .

The Act does not define the phrase "solid waste management purposes" but does define its constituent parts. "[S]olid waste" means "discarded material", including "rubbish", which is "nonputrescible solid waste . . . that consists of . . . combustible waste materials, including paper . . . [and] plastics". "'Management' means the systematic control of the activities of generation, source separation, collection, handling, storage, transportation, processing, treatment, recovery, or disposal of solid waste." The term "management" thus refers to institutional controls imposed at any point in the solid waste stream, from generation of solid waste to disposal.

The Ordinance's stated purpose and its intended effect are to control the generation of solid waste by reducing a source of solid waste on the front end so those single use materials cannot be inappropriately discarded on the back end.

. . .

We think it clear that the Ordinance was adopted for solid waste management purposes.

. . .

In the City's view, the Act does not clearly apply to new bags for point of sale purchases because the term "bag" is not used in the statute and the statute is contextually focused on trash, not new items. As the City sees it, no matter how likely or expeditiously single use bags are destined to become trash, the Act's reach is limited to either (1) containers and packages that have already been discarded, or (2) containers and packages that store or transport garbage, like dumpsters. Again, the City's narrow construction is not supported by a plain reading of the statute.

. . .

The only reasonable construction of the Act that accords with the statute as a whole is one that affords the terms container and package their ordinary meanings.

. . .

Finally, the City argues that the Ordinance escapes preemption because it is "authorized by state law" as shown by its consistency with various state general laws [of which the city presents six major examples].

. . .

But the Act preempts local regulation "in a manner not authorized by state law". The question is not whether a municipality has the power to regulate. Home rule cities already have the power of self governance unless restricted by state law. If "authorized by law" in the preemption provision referred only to the power municipalities already have, the restriction would have no effect. But the preemption provision applies to local regulation when the manner is not authorized by state law. Manner is how something can be done, not merely if it can be. A manner must be stated by, and not merely implied from, a grant of authority. The clear, stated intent of the Act is to control the manner of regulating the sale or use of containers or packages for solid waste management purposes. To conclude otherwise would render the statute meaningless.

. . .

The court of appeals correctly held that the Act preempts the City's Ordinance.

**Source:** City of Laredo, Texas, Petitioner, v. Laredo Merchants Association (NO. 16 0748). In the Supreme Court of Texas. http://www.txcourts.gov/media/1441865/160748.pdf.

### Plastic Bag Bill (Connecticut) (2019); Plastic Bag Bill (Idaho) (2016)

*One of the most popular ways of dealing with plastic wastes in the United States has been the adoption of bills by state legislators that attempt to reduce or end the use of such bags by grocery and other types of stores. In some cases, those bills call for an outright ban on plastic bags. In other cases, they set a fee for the use of plastic bags. The two bills cited here show different approaches to the control of plastic bag use. The Connecticut bill is designed to cut back on plastic bag use, while the Idaho bill takes a somewhat different position. It prohibits individual cities, counties, or other governmental units from passing plastic bag bans, limiting that type of action to decisions by the state legislature only.*

#### Plastic Bag Bill (Connecticut) (2019)
*House Bill 7424*
*[The bill begins with definitions of terms used in the text.]*

(b) (1) For the period commencing August 1, 2019, and ending June 30, 2021, each store shall charge a fee of ten cents for each single use checkout bag provided to a customer at the point of sale. The store shall indicate the number of single use check-out bags provided and the total amount of the fee charged on any transaction receipt provided to a customer. Any fees collected pursuant to this subsection shall be excluded from gross receipts under chapter 219 of the general statutes.

(2) Each store shall report all fees collected pursuant to subdivision (1) of this subsection to the Commissioner of Revenue Services with its return due under section 12 414 of the general statutes and remit payment at the same time and in the same form and manner required under section 12 414 of the general statutes.

(3) Any fees due and unpaid under this subsection shall be subject to the penalties and interest established under section 12 419 of the general statutes and the amount of such fee, penalty or interest, due and unpaid, may be collected under the provisions of section 12 35 of the general statutes as if they were taxes due to the state.

(4) The provisions of sections 12 415, 12 416 and 12 421 to 12 428, inclusive, of the general statutes shall apply to the provisions of this section in the same manner and with the same force and effect as if the language of said sections had been incorporated in full into this section and had expressly referred to the fee imposed under this section, except to the extent that any such provision is inconsistent with a provision of this section.

(5) The Commissioner of Revenue Services, in consultation with the Commissioner of Energy and Environmental Protection, may adopt regulations in accordance with the provisions of chapter 54 of the general statutes, to carry out the provisions of this section.

(6) At the close of each of the fiscal years ending June 30, 2020, and June 30, 2021, the Comptroller is authorized to record as revenue for such fiscal year the amount of the fee imposed under the provisions of this section that is received by the Commissioner of Revenue Services not later than five business days from the last day of July immediately following the end of such fiscal year.

(c) On and after July 1, 2021, no owner or operator of a store shall provide or sell a single use checkout bag to a customer.

(d) Nothing in this section shall be construed to prohibit a municipality from enacting or enforcing an ordinance concerning single use checkout bags made of plastic, provided such ordinance is as restrictive or more restrictive as the provisions of this section concerning the provision or selling of such bags to customers by stores. Nothing in this section shall be construed to prohibit a municipality from enacting or enforcing an ordinance concerning single use checkout bags made of paper, including, but not limited to, enabling each store to charge a fee for any such bag distributed to a customer.

**Source:** House Bill No. 7424. 2019. Connecticut General Assembly. https://www.cga.ct.gov/2019/ACT/pa/pdf/2019PA-00117-R00HB-07424-PA.pdf.

### Plastic Bag Bill (Idaho) (2016)

SECTION 1. LEGISLATIVE INTENT. It is the intent of the Legislature that prudent regulation of auxiliary containers is crucial to the welfare of Idaho's economy; that retail and food establishments are sensitive to the costs and regulation of auxiliary containers; and, that if individual political subdivisions of the state regulate such auxiliary containers, there exists the potential for confusing and varying regulations which could lead to unnecessary increased costs for retail and food establishments to comply with such regulations.

SECTION 2. That Chapter 23, Title 67, Idaho Code, be, and the same is hereby amended by the addition thereto of a NEW SECTION, to be known and designated as Section 67 2340, Idaho Code, and to read as follows:

67 2340. REGULATION OF AUXILIARY CONTAINERS. (1) As used in this section, "auxiliary container" means reusable bags, disposable bags, boxes, cups and bottles which are made of cloth, paper, plastic, extruded polystyrene or similar materials that are designed for one time use or for transporting merchandise or food from food and retail facilities.

(2) Any regulation regarding the use, disposition or sale or any imposition of any prohibition, restriction, fee imposition or taxation of auxiliary containers at the retail, manufacturer

or distributor setting shall be imposed only by statute enacted by the legislature.

(3) Nothing in this section shall be construed to prohibit or limit any county or municipal curbside recycling program or other designated residential or commercial recycling location.

(4) The provisions of subsection (2) of this section shall not apply to the use of auxiliary containers in any event organized, sponsored or permitted by a county, municipality or school district on a property owned by such county, municipality or school district.

**Source:** House Bill No. 372. 2016. Legislature of the State of Idaho. https://legislature.idaho.gov/wp content/uploads/session info/2016/legislation/H0372.pdf.

**A Circular Economy for Plastics (2019)**

*In 2019, the European Union released a report dealing with the problem of plastic pollution in the European Union. Written by a group of experts in the field, the report included sections on Plastic Pollution, Substances of Concern to Human and Environmental Health, New Materials, Biological Feedstock, Collection and Sorting, Chemical Recycling, and Organic Recycling and Biodegradation. Based on information gathered for the report, the authors offered several recommendations for dealing with the problem of plastic pollution, as reprinted here. (This report overall is an extremely valuable summary of the plastic pollution problem and efforts that can be made to deal with the problem.)*

SHORTLIST OF POLICY RECOMMENDATIONS

The list below represents a high level synthesis of recommendations proposed by the experts and edited following feedback from a wider stakeholder group. More details and the underlying reasoning can be found in the different chapters and in APPENDIX: Overview policy recommendations.

General insights across the plastics value chains

1. Facilitate collaboration across the plastics value chains towards a common vision to trigger actions on a regional, national, European and global level.

2. Develop, harmonise and enforce regulatory and legal frameworks guided by systems thinking to connect the different actors of the plastics value chain(s).

3. Set up, connect and fund mechanisms to coordinate strategically the transition towards a circular economy and to invest in upstream and downstream capacity across Europe.

4. Provide funding for research and a range of financial incentives for systemic innovation in business models, products and materials fit for a circular economy for plastics.

5. Educate and support citizens, companies and investors on the transition towards a circular economy for plastics.

Part I: The unintended impacts of plastics on society and the environment

6. Harmonise definitions, frameworks for data gathering and analyses of plastic pollution sources, pathways, fates and impacts at a European and global level.

7. Develop open collaboration platforms to enable more comprehensive analyses and frequent benchmarking on plastic flows and impacts, to provide information on and for investments, and to inform industry, government and the public.

8. Enforce, harmonise and adapt existing EU chemical regulations (e.g. REACH, Toy Safety Directive, regulation on food contact materials) based on a systems thinking approach.

9. Develop regulatory frameworks with additional requirements for additives and other chemicals in plastic products based on the overall migrate and the potential toxicity of the mixture from combined exposure to finished articles.

10. Provide business support to identify and reduce chemical hazards, and to create transparency on the socio economic and environmental impacts of plastics and on successful alternative solutions.

Part II: Novel sources, designs and business models for plastics in a circular economy

11. Facilitate gathering, sharing and trading of reliable information and data on business models, technologies and material composition to foster open innovation and activation of industry, government, innovators and the public.

12. Set up a coordination mechanism, combining technical, commercial and behavioural expertise, for tracking material flows and renewable feedstock inventories, and for strategic long term investments in plastics production, collection, sorting and recycling infrastructure across Europe.

13. Develop regulatory measures such as standards, assessment methodologies, ecodesign requirements and incentives such as Extended Producer Responsibility (EPR) schemes with modulated fees, to evaluate and steer design of business models and products towards elimination of challenging items, use of renewable or recycled feedstock, reuse and cost effective recycling, and to fund innovation in this field (e.g. through Packaging and Packaging Waste Directive (PPWD), Ecodesign Directive, Waste Framework Directive (WFD)).

14. Set up, connect and participate as an active stakeholder or shareholder in investment instruments to enable investors and lenders to provide funds for circular economy business models (Horizon Europe).

15. Provide regulatory, legal and financial incentives to support long term R&I in chemicals and materials based on renewable feedstock and recycled materials, and their scale up towards a self sustaining critical mass, while ensuring

environmental benefits based on a holistic impact assessment across the life cycle.

16. Provide information for citizens and businesses about materials based on renewable feedstock and about recycled materials by developing standards, labels and a holistic impact assessment framework.

17. Incorporate systems thinking and circular design in the education curriculum at all levels.

Part III: Circular after use pathways for plastics

18. Develop a holistic vision for an after use plastics system in Europe, incorporating reuse and repair, and mechanical, chemical and organic recycling, and develop a methodology for comparing these different options based on feasibility, and on the environmental, economic and social impact.

19. Facilitate gathering and sharing of reliable information and data on virgin and recycled material composition and on collection, sorting and recycling performance and best practice cases, to enable cross value chain collaboration and compatibility.

20. Develop a regulatory framework to harmonise collection systems, allowing a certain degree of local adaptation to socio economic conditions.

21. Develop regulatory measures, such as ecodesign requirements, and financial incentives, such as EPR with modulated fees, integrating new digital technologies, to evaluate and steer design of business models and products towards elimination of challenging items, use of renewable or recycled feedstock, reuse and cost effective recycling, and to fund innovation in this field (e.g. through PPWD, Ecodesign Directive and WFD).

22. Develop and implement harmonised standards for the quality of mechanically and chemically recycled plastics,

and for verification of recycled content, taking into account safety and application areas.

23. Provide regulatory and fiscal incentives to stimulate the demand for recycled plastics, such as public procurement, and to take into account the costs of negative externalities associated with different feedstock types, such as reduced value added tax (VAT).

24. Review and update waste legislation to incorporate the latest recycling technologies, including end of waste criteria for plastics, guided by systems thinking and the European strategy for plastics in a circular economy.

25. Harmonise regulatory efforts, including standardisation, to provide direction for R&I and implementation of compostable and biodegradable materials, and to establish clear communication and guidance for citizens and business.

**Source:** Cripo, Maurizio, et al. 2019. "A Circular Economy for Plastics." Brussels, Belgium: European Commission, 11–12. https://op.europa.eu/en/publication detail//publication /33251cf9 3b0b 11e9 8d04 01aa75ed71a1/language en/format PDF/source 87705298 (site discontinued).

### Save Our Seas Bill (S.1982) (2019)

*By the end of the second decade of the twenty-first century, politicians had become aware of the need to develop stronger legislation to deal with the problem of plastic wastes, especially as they affect the oceans. In 2018 and 2020, two major bills were submitted to address problems of plastic wastes in the seas. The first of these bills passed the Congress and was signed into law by President Donald Trump on October 11, 2018. The second bill had not been acted upon by either House as of mid-2020. Relevant sections of each bill are reprinted here.*

SEC. 133. INCENTIVE FOR FISHERMEN TO COLLECT AND DISPOSE OF PLASTIC FOUND AT SEA.

(a) IN GENERAL.-The Under Secretary shall establish a pilot program to assess the feasibility and advisability of providing incentives, such as grants, to fishermen based in the United States who incidentally capture 21 marine debris while at sea-

(1) to track or keep the debris on board; and

(2) to dispose of the debris properly on land.

(b) SUPPORT FOR COLLECTION AND REMOVAL OF DERELICT GEAR.—The Under Secretary shall encourage United States efforts, such as the Fishing for Energy net disposal program, that support-

(1) collection and removal of derelict fishing gear and other fishing waste;

(2) disposal or recycling of such gear and waste; and

(3) prevention of the loss of such gear.

. . .

SEC. 141. REPORT ON OPPORTUNITIES FOR INNOVATIVE 6 USES OF PLASTIC WASTE.

Not later than 2 years after the date of enactment of this Act, the Interagency Marine Debris Coordinating Committee shall submit to Congress a report on innovative uses for plastic waste in consumer products.

SEC. 142. REPORT ON MICROFIBER POLLUTION.

Not later than 2 years after the date of the enactment of this Act, the Interagency Marine Debris Coordinating Committee shall submit to Congress a report on microfiber pollution that includes—

(1) a definition for "microfiber";

(2) an assessment of the sources, prevalence, and causes of microfiber pollution;

(3) a recommendation for a standardized methodology to measure and estimate the prevalence of microfiber pollution;

(4) recommendations for reducing microfiber pollution; and

(5) a plan for how Federal agencies, in partnership with other stakeholders, can lead on opportunities to reduce microfiber

pollution during the 5 year period beginning on such date of enactment.

SEC. 143. STUDY ON UNITED STATES PLASTIC POLLUTION DATA.

(a) IN GENERAL.—The Under Secretary, in consultation with the EPA Administrator and the Secretary of the Interior, shall seek to enter into an arrangement with the National Academies of Sciences, Engineering, and Medicine under which the National Academies will undertake a multifaceted study that includes the following:

(1) An evaluation of United States contributions to global ocean plastic waste, including types, sources, and geographic variations.

(2) An assessment of the prevalence of marine debris and mismanaged plastic waste in saltwater and freshwater United States navigable waterways and tributaries.

(3) An examination of the import and export of plastic waste to and from the United States, including the destinations of the exported plastic waste and the waste management infrastructure and environmental conditions of these locations.

(4) Potential means to reduce United States contributions to global ocean plastic waste.

. . .

SEC. 202. PRIORITIZATION OF EFFORTS AND ASSISTANCE TO COMBAT MARINE DEBRIS AND IMPROVE PLASTIC WASTE MANAGEMENT.

(a) IN GENERAL.—The Secretary of State shall, in coordination with the Administrator of the United States Agency for International Development, as appropriate, and the officials specified in subsection (b)—

(1) lead and coordinate efforts to implement the policy described in section 201; and

(2) develop strategies and implement programs that prioritize engagement and cooperation with foreign governments, subnational and local stakeholders, and the private sector to expedite efforts and assistance in foreign countries-

(A) to partner with, encourage, advise and facilitate national and subnational governments on the development and execution, where practicable, of national projects, programs and initiatives to—

(i) improve the capacity, security, and standards of operations of post consumer materials management systems;

(ii) monitor and track how well postconsumer materials management systems are functioning nationwide, based on uni9 form and transparent standards developed in cooperation with municipal, industrial, and civil society stakeholders;

(iii) identify the operational challenges of post consumer materials management systems and develop policy and programmatic solutions;

(iv) end intentional or unintentional incentives for municipalities, industries, and individuals to improperly dispose of plastic waste; and

(v) conduct outreach campaigns to raise public awareness of the importance of proper waste disposal and the reduction of plastic waste;

(B) to facilitate the involvement of municipalities and industries in improving solid waste reduction, collection, disposal, and reuse and recycling projects, programs, and initiatives;

(C) to partner with and provide technical assistance to investors, and national and local institutions, including private sector actors, to develop new business opportunities and solutions to specifically reduce plastic waste and expand solid waste and post consumer materials management best practices in foreign countries by—

(i) maximizing the number of people and businesses, in both rural and urban communities, receiving reliable solid waste and post consumer materials management services;

(ii) improving and expanding the capacity of foreign industries to responsibly employ post consumer materials management practices;

(iii) improving and expanding the capacity and transparency of tracking mechanisms for marine debris to reduce the impacts on the marine environment;

(iv) eliminating incentives that undermine responsible post consumer materials management practices and lead to improper waste disposal practices and leakage;

(v) building the capacity of countries-

(I) to reduce, monitor, regulate, and manage waste, post consumer materials and plastic waste, and pollution appropriately and transparently, including imports of plastic waste from the United States and other countries;

(II) to encourage private investment in post consumer materials management and reduction; and

(III) to encourage private investment, grow opportunities, and develop markets for recyclable, reusable, and repurposed plastic waste and post consumer materials, and products with high levels of recycled plastic content, at both national and local levels; and

(vi) promoting safe and affordable reusable alternatives to disposable plastic products, to the extent practicable; and

(D) to research, identify, and facilitate opportunities to promote collection and proper disposal of damaged or derelict fishing gear.

**Source:** S. 1982. 2019. Congress.gov. https://www.congress .gov/bill/116th-congress/senate-bill/1982.

### Plastic or Bioplastic Disposable Food Service Ware (2019)

*Many cities and states have attempted to deal with pollution from plastic wastes by passing ordinance that prohibits the use of plastic straws, bags, or other objects. Few municipalities have gone so far, however, as to ban all types of plastic objects used in restaurants or similar settings. One such governmental body has been the city of Talent, Oregon, which, in 2019, adopted an ordinance to ban the use of all plastic-based food service materials. The ordinance took effect in January 2020. Important parts of the ordinance are as follows.*

8.27.010
Intent.

It is the intent of the city of Talent to prohibit the distribution of disposable food service ware derived from plastic or bioplastic by retail establishments and others operating within the city limits by the following regulations. [Ord. 957, 2019.]

8.27.020
Definitions.

"Alternative" means nonplastic substitutions for disposable service ware such as those made from paper, hemp, bamboo, wheat, seaweed, fungi, glass, sugar cane, or pasta.

"Bioplastic" means a plastic like unit made substantially from renewable plant materials such as cellulose or starch designed to look and act like plastic, but which contaminates the traditional recycling stream.

"City manager" means the city manager of the city of Talent or the city manager's designee.

"City hosted event" means any event held on city of Talent property and that is the subject of a rental agreement with the city.

"City sponsored event" means any event organized or officially sponsored by the city of Talent or any department of the city of Talent.

"Disposable service ware" means all containers, bowls, plates, cups, lids, stirrers, cutlery, hinged or lidded containers (clamshells) and other like items that are designed for one time use for prepared foods, including, without limitation, service ware for takeout foods and/or leftovers from partially consumed meals prepared by food vendors.

"Food vendor" means, but is not limited to, shops, sales outlets, restaurants, bars, pubs, coffee shops, cafeterias, caterers, convenience stores, liquor stores, grocery stores, supermarkets, delicatessens, nonprofit organizations, mobile food trucks, vehicles or carts, and roadside stands that serve, sell, or otherwise provide food or beverages for human consumption.

"Plastic" means a synthetic material produced by the synthesis from primary chemicals generally coming from oil, natural gas, or coal. "Plastic" means, but is not limited to, polyethylene, high density polyethylene, low density polyethylene, polyethylene terephthalate, polypropylene, and polyvinylidene chloride.

"Prepared food" means, but is not limited to, food or beverages that are packaged, cooked, chopped, sliced, mixed, brewed, frozen, squeezed, or otherwise prepared on the premises. "Prepared foods" does not mean: (1) any raw meat product unless it can be consumed without any further preparation, or (2) prepackaged food that is delivered to the food vendor wholly encased, contained, or packaged in a container or wrapper, and sold or otherwise provided by the food vendor in the same container or packaging. It is recommended that the food vendor communicate the need to reduce plastic packaging and engage in identifying alternatives.

"Provide" means, but is not limited to, active serving, giving away, selling, delivering, packaging, and providing. [Ord. 957 § 1, 2019.]

8.27.030
Distribution of disposable food service ware derived from plastic or bioplastic prohibited.

A. No food vendor shall provide, distribute, or sell individual items of disposable service ware made of plastic or bioplastic to its customers.

B. Nothing in this section precludes food vendors from using or making alternatives available to customers. Alternatives to items such as stirrers and cutlery shall only be provided upon request by the customer. [Ord. 957 § 2, 2019.]

8.27.040
Use or distribution prohibited during city events.

No person(s) shall use or provide disposable service ware made of plastic or bioplastic during any city sponsored, city permitted, or city hosted event at any city facility or city managed concession.

A. An event organizer may request an exemption from the city manager where neither a durable replacement nor an alternative is available, such as lids and cups for hot beverages. Exemption requests must be made in writing at least 14 days prior to the event date. All decisions by the city manager are final. [Ord. 957 § 3, 2019.]

8.27.050
Other offerings.

A. A food vendor may alternatively offer disposable service ware made of paper, hemp, bamboo, wheat, seaweed, fungi, glass, sugar cane, or pasta at the customer's request.

B. A food vendor may make nondisposable food service ware available for use on site, for an additional charge, or other appropriate collateral.

C. A food vendor may offer a discount to a customer who brings their own durable food service ware. [Ord. 957 § 4, 2019.]

**Source:** "Chapter 8.27. Plastic or Bioplastic Disposable Food Service Ware." 2019. City of Talent. https://talent.municipal .codes/TMC/8.27.

### San Antonio Bay Estuarine Waterkeeper and S. Diane Wilson v. Formosa Plastics Corp., Texas, and Formosa Plastics Corp., U.S.A. (2019)

*Diane Wilson is a retired shrimper whose family has lived on the Central Texas coast for more than a century. Over the past few decades, she and her neighbors have seen an accumulation of plastic wastes in the waterways surrounding their homes. The primary culprit in this problem appeared to have been the Formosa Plastics Corporation, owners and operators of a petrochemical plant headquartered in Taiwan. When Formosa declined to make the changes in their operating procedures that Wilson and her colleagues believed were necessary to protect their environment from plastic and microplastic wastes, she brought suit in the United States District Court for the Southern District of Texas, Victoria Division. After hearing the case and reviewing negotiations between the parties, District Judge Kenneth M. Hoyt approved a settlement that largely included remedies for the problems that Wilson and her co-plaintiffs had raised. The following contains excerpts from the long and detailed agreement reached in the case.*

IV. REMEDIAL MEASURES

A. Engineering Changes

Within twenty one (21) Days of the effective date of this Consent Decree, Formosa will propose to Plaintiffs an Engineering Consultant to review the current design and operation of the Formosa Point Comfort Plant with regard to the discharge of Plastics and, ultimately, to design and audit the effectiveness of measures to halt those discharges.

. . .

The Engineering Consultant shall, consistent with good engineering principles, produce plans to retrofit the facility with the Best Available Technology and design to prevent the discharge of Plastics, including the following plans to address deficiencies in Formosa's current system:

a. capacity improvements to the stormwater drainage system such that flooding does not occur from rainfall that is, at least, a 5 year 24 hour rainfall event (6.8 inches in 24 hours);

. . .

b. direction of all stormwater outside battery limits for Outfalls 002, 003, 004, 005, 006, 007, 008, 009, 012 and 014 to a holding pond system that is designed to have zero (0) discharge of stormwater into Cox Creek for at least a 5 year 24 hour rainfall event (6.8 inches in 24 hours). The ponds and stormwater ditches will include engineered controls to remove Plastics if stormwater will be discharged into Cox Creek in the event of a greater than 5 year 24 hour rainfall event (6.8 inches in 24 hours);

c. improvements to the inside boundary limits stormwater and wastewater systems to remove Plastics prior to entry into the current combined wastewater treatment plant ("CWTP") system; and

d. improvements in source reduction at all manufacturing and loading units.

. . .

B. Monitoring, Reporting, and Future Mitigation Payments

Within thirty (30) Days of the effective date of this Consent Decree, the Plaintiffs and Formosa will each propose two (2) recommendations for a Monitor.

. . .

At least twice per week, the Monitor will examine (a) the outfalls, containment Booms, water and adjacent shores fifty (50) feet downstream and upstream of containment Booms for all Formosa outfalls discharging into Cox Creek and (b) the Bypass Pipe WSM.

a. Subject to each individual seeking access executing a full release and indemnification of Formosa from any and all liability resulting from access, Plaintiffs will have unrestricted access to all discharge channels and the shore areas upstream of the Booms outside Formosa's fence line for outfalls that discharge into Cox Creek.

b. If Formosa or its subcontractors detect Plastics upstream of a Boom or the shore areas upstream of the Booms, Formosa or its subcontractor will take photographs of the Plastics detected and will report that information to the Monitor and Remediation Consultant within twenty four (24) hours.

c. Booms shall not be modified or moved without prior notice to Plaintiffs and the Monitor. Plaintiffs and the Monitor will have ten (10) Days from receipt of notice to object to the modification of Booms and that objection may be appealed to the Court if the issue cannot be resolved between Plaintiffs and Formosa. For purposes of this subparagraph, the phrase "modified or moved" does not include routine maintenance and repair of the Booms.

d. Within thirty (30) Days of the effective date of this Consent Decree, Formosa shall install and/or maintain Booms for all outfalls discharging to Cox Creek, including Outfalls 003, 007, 010, and 012.

. . .

C. Remediation of Past Discharges

Within thirty (30) Days of the effective date of this Consent Decree, the Plaintiffs and Formosa will each propose two (2) recommendations for a Remediation Consultant.

. . .

The Remediation Consultant will review remediation methods for Plastics in Cox Creek and Lavaca Bay as follows:

a. The Remediation Consultant will review ongoing remediation methods, make site visits, and will produce a Remediation Plan with the goal being a removal of most Plastics from the environment while protecting the Cox Creek and Lavaca Bay ecosystems.

b. Formosa will direct Horizon Environmental Services, Inc. ("Horizon") to provide whatever information regarding previous cleanup efforts in Horizon's possession as is requested by the Remediation Consultant.

c. The Remediation Consultant will determine whether Horizon's removal of shoreline and submerged vegetation should be remedied by planting of native vegetation to stabilize the banks and what mitigation should be undertaken, if any, in Cox Creek where submerged vegetation has been removed. Formosa will cease any efforts to authorize removal of submerged vegetation from Cox Creek or its shoreline at Texas Parks and Wildlife Department, and Formosa will withdraw its request within ten (10) Days of the effective date of this Consent Decree.

d. If the Remediation Consultant determines specific removal methods will cause significant environmental damage that cannot be remediated, the removal methods should not be used.

e. Plaintiffs may request the opportunity to take the Remediation Consultant up Cox Creek or to Lavaca Bay to show areas of concern. Dr. Jeremy Conkle may accompany Plaintiffs and the Remediation Consultant during this process, or

Dr. Conkle may speak directly with the Remediation Consultant to discuss Dr. Conkle's concerns.

. . .

D. Permit and Mitigation Terms

Within ten (10) Days of the effective date of this Consent Decree, Formosa will request in writing to TCEQ as part of the current Texas Pollutant Discharge Elimination System ("TPDES") permit renewal process that Formosa's renewed TPDES permit contain the following new permit requirements:

a. There will be zero (0) discharge of stormwater or other waters, including washwater, from Outfalls 002, 003, 004, 005, 006, 007, 008, 009, 012 and 014 for rainfall events of 5 year 24 hour rainfall event or less (6.8 inches in 24 hours) as measured by Formosa's onsite rain gauge. If Formosa discharges stormwater or other waters from these Outfalls, it will notify TCEQ within twenty four (24) hours of the discharge and include the rainfall amount and outfall number. This requirement will be effective upon the earlier of actual construction of the infrastructure necessary to achieve this requirement or January 1, 2024.

b. There will be zero (0) discharge of Plastics from Formosa's Point Comfort Plant. Formosa shall not propose that Plastics in its discharge in any way are part of permitted total suspended solids.

c. Formosa will provide Plaintiffs a copy of the letter to TCEQ complying with this term immediately after sending it. Formosa will also provide Plaintiffs copies of any subsequent correspondence and notifications of in person or oral communications to or from TCEQ regarding these proposed permit terms within five (5) Days.

d. Regardless of whether TCEQ adopts these new permit terms proposed by Formosa, Formosa agrees to comply with the requirements in paragraphs 49 (b) and (c) above, on the effective date of this Consent Decree, and with 49(a) upon the earlier of actual construction of the infrastructure necessary to achieve this requirement or January 1, 2024.

. . .

E. Environmental Mitigation Projects

Establishment of the Matagorda Bay Mitigation Trust

Formosa will pay fifty (50) million dollars over five (5) years (ten (10) million dollars per year) for Mitigation Projects to the Matagorda Bay Mitigation Trust.

. . .

Environmental Mitigation Projects

The total amount for Mitigation Projects will be divided as follows:

Twenty (20) million dollars to the Federation of Southern Cooperatives ("the Federation"), a non profit organization with offices throughout the South, to form a Matagorda Bay Fishing Cooperative ("the Cooperative"), and netting or transportation cooperatives if necessary to support the Fishing Cooperative, under a project called the Matagorda Bay Cooperative Development Project ("the Project").

[Details of each of the monetary awards are discussed in sections following each award.]

. . .

Ten (10) million dollars in total for the development, protection, operation and maintenance of Green Lake Park.

. . .

Seven hundred fifty (750) thousand dollars to the Port Lavaca YMCA to fund camps for children and teenagers in the area, which will be focused on education about how to be a good steward of the local ecosystems and will teach outdoor education and recreation activities.

. . .

Two (2) million dollars to Calhoun County for erosion control and beach restoration at Magnolia Beach. Funds may be used for purchase and use of clean and uncontaminated fill material, planting of native plants, necessary construction to prevent future erosion, and necessary maintenance to prevent beach erosion.

. . .

One (1) million dollars to the University of Texas Marine Science Institute ("UTMSI") Nurdle Patrol ("the Nurdle Patrol"). The money will be used to support the Nurdle Patrol and for workshops and meetings, and to provide scholarships for attendance, food, transportation and expenses at conferences.

. . .

Five (5) million dollars for an Environmental Research Mitigation Project providing for funding for environmental research regarding the Bay Systems, or the river deltas in Calhoun or Jackson Counties feeding into those systems. Funding may be distributed for environmental research topics, including but not limited to, the ecology, pollution, fisheries, or habitat and wildlife restoration of the ecosystems. The Trustee shall set up a system to provide funding for research, including providing public notice of the funding opportunity and setting up a review process for applications with researchers in applicable fields.

. . .

Eleven (11) million two hundred and fifty (250) thousand dollars to the Matagorda Bay Mitigation Trust, and any additional sums paid by Formosa or returned to the Trust for redistribution.

**Source:** San Antonio Bay Estuarine Waterkeeper and S. Diane Wilson Vs. Formosa Plastics Corp., Texas, and Formosa Plastics Corp., U.S.A. 2019. https://static1.squarespace.com/static /5b58f65a96d455e767cf70d4/t/5de5306c5fb6fb30a7bdd7dc /1575301231792/Final+consent+decree.pdf.

### Udall-Lowenthal Bill (2020)

*This bill was originally written in 2019, then reviewed and commented upon over a period of months before being formally filed in February 2020. The following is a general summary of the bill provided by its sponsors in an email to contributors and commentators in late January 2020 and a final draft of the bill as submitted to the Congress.*

Attached is a revised and updated discussion draft of the Udall/ Lowenthal proposal along with a redline copy to demonstrate the hundreds of changes and edits that were made to reflect this extensive stakeholder process.

Require Plastic Producers to Take Responsibility for Collecting and Recycling Materials: Producers of covered products will be required to design, manage, and finance programs to collect and process waste that would normally burden state and local governments. The legislation will encourage producers to cooperate with those who produce similar products to take responsibility for their waste and implement cleanup programs with Environmental Protection Agency approval.

Producers will cover the costs of waste management and clean up, as well as awareness raising measures for covered products, which includes packaging, containers, paper, and food service products, regardless of the recyclability, compostability, and type of material (including plastic, paper, glass, metal, etc.).

Require Nationwide Container Refunds: The legislation will institute a 10 cent national refund requirement for all beverage containers, regardless of material, to be refunded to customers when they return containers. Any unclaimed refunds will go to beverage producers to offset investments in nationwide collection and recycling infrastructure. This legislation encourages states that have already implemented similar initiatives to continue their current systems if they match the federal requirements.

Phase Out Certain Polluting Products: Beginning in January 2022, some of the most common single use plastic products that pollute our environment and cannot be recycled will be phased out from sale and distribution. The prohibitions will apply to lightweight plastic carryout bags, food and drinkware from expanded polystyrene, and plastic utensils.

Carryout Bag Fee: The legislation would impose a fee on the distribution of carryout bags. The fee can be retained by retailers who implement a reusable bag credit program at their place of business. For fees collected by those who do not participate in a reusable bag credit program, the fee will be used to fund access to reusable bags as well as litter clean up and recycling infrastructure.

Minimum Recycled Content Requirement: Plastic beverage containers will be required to include an increasing percentage of recycled content in their manufacture before entering the market. Additionally, the EPA will be required to implement post consumer minimum recycled content for other covered products and material types after a review with the National Institute of Standards and Technology is completed to determine technical feasibility.

Recycling and Composting: The EPA will develop standardized recycling and composting labels for products and receptacles to encourage proper sorting and disposal of items that can be recycled or composted.

Plastic Tobacco Filters, Electronic Cigarettes and Derelict Fishing Gear: Following studies on the environmental impacts of plastic tobacco filters, electronic cigarette parts and derelict fishing gear, the relevant agencies will propose measures to reduce those environmental impacts.

Prevent Plastic Waste from Being Shipped to Developing Countries that Cannot Manage It: The bill will prevent the export of plastic waste, scrap and pairings to non OECD countries, many of whom have been a major source of ocean plastic pollution due to their inability to manage the waste.

Protect Existing State Action: The bill would protect state and local governments to enact more stringent standards, requirements, and additional product bans.

Temporary Pause on New Plastic Facilities: The legislation will give environmental agencies the valuable time needed to investigate the cumulative impacts of new and expanded plastic producing facilities on the air, water, climate, and communities before issuing new permits to increase plastic production. The legislation would also update EPA regulations to eliminate factory produced plastic pollution in waterways and direct the EPA to update existing Clean Air and Clean Water Act emission and discharge standards to ensure that plastic producing facilities integrate the latest technology to prevent further pollution.

. . .

[Some basic parts of certain critical sections read as follows:]
PART I PRODUCTS IN THE MARKETPLACE
SEC.12101.EXTENDEDPRODUCERRESPONSIBILITY.

(a) In General. Except as provided in subsection (b), beginning on February 1, 2023, each responsible party for any covered product or beverage sold in a beverage container that is sold, distributed, or imported into the United States shall

(1) participate as a member of an Organization for which a Plan is approved by the Administrator; and

(2) through that participation, satisfy the performance targets under section 12105(g).

(b) Exemptions. A responsible party for a covered product or beverage sold in a beverage container, including a responsible party that operates as a single point of retail sale and is not supplied by, or operated as part of, a franchise, shall not be subject to this part if the responsible party

. . .

SEC. 12104. NATIONAL BEVERAGE CONTAINER PROGRAM.

(a) Responsibilities of Responsible Parties.

(1) IN GENERAL. Each responsible party for beverages sold in beverage containers shall

(A) charge to a retailer to which the beverage in a beverage container is delivered a deposit in the amount of the applicable refund value described in subsection (c) on delivery; and

(B) on receipt of an empty beverage container from a retailer, pay to the retailer a refund in the amount of the applicable refund value described in subsection c.

(2) USE OF DEPOSITS FROM UNREDEEMED BEVERAGE CONTAINERS. A responsible party shall use any amounts received as deposits under paragraph (1)(A) for which an empty beverage container is not returned to the Organization responsible for the material of the beverage

container for investment in collection, recycling, and reuse infrastructure.

. . .

SEC. 12201. PROHIBITION ON SINGLE USE PLASTIC CARRYOUT BAGS.

(a) Definition of Single use Plastic Bag. In this section:

(1) IN GENERAL. The term "single use plastic bag" means a bag that is

(A) made of

(i) plastic film; or

(ii) woven or nonwoven nylon, polypropylene, polyethylene terephthalate, or Tyvek in a quantity less than 80 grams per square meter; and

(B) provided by a covered retail or service establishment to a customer at the point of sale, home delivery, the check stand, cash register, or other point of departure to a customer for use to transport, deliver, or carry away purchases.

(2) EXCLUSIONS. The term "single use plastic bag" does not include

[Several exclusions are listed here, such as a bag used by a customer within a store, a newspaper bag, or a laundry or dry cleaning bag.]

. . .

SEC. 12202. REDUCTION OF OTHER SINGLE USE PRODUCTS.

(a) Prohibition on Plastic Utensils and Plastic Straws.

(1) UTENSILS. A covered retail or service establishment may not use, provide, distribute, or sell a plastic utensil.

(2) PLASTIC STRAWS. A covered retail or service establishment may provide a plastic straw to a customer only on request of the customer.

(3) NONPLASTIC ALTERNATIVES. Nothing in this subsection precludes a covered retail or service establishment from using or making available to a customer, on request of the customer, reusable, compostable, or recyclable alternatives to plastic utensils or plastic straws, such as alternatives made

from aluminum, paper, grain stalks, sugar cane, or bamboo, that are supplied by a responsible party that participates in an Organization.

(b) Prohibition on Other Single use Products.

(1) IN GENERAL. Except as provided in paragraphs (3) and (4), a covered retail or service establishment may not sell or distribute any single use product that the Administrator determines is not recyclable or compostable and can be replaced by a reusable or refillable item.

(2) INCLUSIONS. In the prohibition under paragraph (1), the Administrator shall include

(A) expanded polystyrene for use in food service products, disposable consumer coolers, or shipping packaging;

(B) single use personal care products, such as miniature bottles containing shampoo, soap, and lotion that are provided at hotels;

(C) noncompostable produce stickers; and

(D) such other products that the Administrator determines by regulation to be appropriate.

(3) EXCEPTION. The prohibition under paragraph (1) shall not apply to the sale or distribution of an expanded polystyrene cooler for medical use.

(4) TEMPORARY WAIVER. The Administrator may grant a temporary waiver of not more than 1 year from the prohibition under paragraph (1) for the use of expanded polystyrene in shipping packaging to protect a product of high value if a viable alternative to expanded polystyrene is not available.

**Source:** Jonathan Black. 2020. "Update on Udall/Lowenthal Plastic Pollution Bill." [Email to interested persons]. Final version may be subject to corrections and modifications of a relatively modest nature. See later email modifications, such as February 13, 2020. Available from bill sponsors.

# 6 Resources

The discovery, invention, and use of plastics, microplastics, nanoplastics, and related materials are among the greatest achievements of modern chemistry. Their availability has transformed many forms of technology, ranging from the packaging of small amounts of medication to the structure of the largest jet airplanes. A price that society has to pay for the many benefits of plastics, however, is its appearance in the waste stream of municipal, industrial, agricultural, and other operations. This interplay of plastic benefits and risks has been the topic of untold numbers of books, articles, reports, and online web pages. This chapter can provide no more than an introduction to some of the items available in one or another of these forms. In some cases, an item may occur in more than one form, such as a print article and an online web page. In such a case, the item is listed first in its print form, with a link to its online form. Readers should also be aware of the wealth of resource ideas contained in the notes at the end of chapters 1 and 2 of this book.

## Books

Abbing, Michael Roscam. 2019. *Plastic Soup: An Atlas of Ocean Pollution*. Washington, DC: Island Press.
> This book is a heavily illustrated discussion of all aspects of the issue of plastic waste pollution in the world's oceans.

---

Gull with its head caught in a plastic six-pack holder. Ocean debris harms wildlife when they become trapped in it or ingest it. (Pancaketom/Dreams time.com)

Al-Salem, Sultan. 2019. *Plastics to Energy: Fuel, Chemicals, and Sustainability Implications*. Kidlington, UK: William Andrew.

This book discusses developments in end-of-life procedures for treating plastic wastes. It discusses features that will be of interest to engineers, material scientists, waste management administrators, and the general public. It also reviews the latest trends in the practical applications of these procedures.

Baur, Erwin, Tim A. Oswald, and Natalie Rudolph. 2019. *Plastics Handbook: The Resource for Plastics Engineers*. 5th ed. Munich, Germany: Hanser Publications.

Although this book is quite technical, designed specifically for specialists in the field, it contains a fair amount of information that will be of interest to the general reader.

Braungart, Michael, and William Donough. 2019. *Cradle to Cradle: Remaking the Way We Make Things*. London: Vintage.

Two of the pioneers in the modern "cradle-to-cradle" movement outline the basic principles and applications of this philosophy for the production, use, and recovery of plastic and other materials. A brief, overall introduction to this topic is available online at https://mcdonough .com/writings/cradle-cradle-alternative/.

Buffington, Jack. 2019. *Peak Plastic: The Rise or Fall of Our Synthetic World*. Santa Barbara, CA: Praeger.

The author begins by pointing out both the benefits and dangers of plastics in our world. He then argues that, to date, the former have outweighed the latter. That situation will change, he suggests, in about 2030, when the availability of plastics will pose more problems than the benefits they provide. He discusses some ways in which this problem can be met and solved.

Clark, Teresa. 2015. *Plastics: Establishing the Path to Zero Waste: A Pragmatic Approach to Sustainable Management of Plastic*

*Materials.* Scotts Valley, CA: CreateSpace Independent Publishing Platform.

The author introduces the basic concepts associated with zero waste management of plastics and discusses biodegradable plastics, environmental marketing of plastics, and a sustainable approach to plastic waste.

De Smet, Michiel, and Mats Linder, eds. 2019. *A Circular Economy for Plastics: Insights from Research and Innovation to Inform Policy and Funding Decisions.* Brussels: European Commission. https://op.europa.eu/en/publication-detail/-/publication /33251cf9-3b0b-11e9-8d04-01aa75ed71a1/language-en /format-PDF/source-87705298.

The concept of a circular economy for plastics is far more than a theoretical idea. It is an approach to plastic wastes that has been studied and tested in some detail. This publication describes several settings and research studies that have examined one or another feature of a circular economy for plastics and suggests ways in which this information can be used to develop real-world circular economy programs.

Endres, Hans-Josef. 2019. "Bioplastics." In Kurt Wagemann and Nils Tippkötter, eds. *Biorefineries.* Cham, Switzerland: Springer Nature, 427–468. Publication: *Advances in Biochemical Engineering/Biotechnology*, 166 (2019): 427–468.

This chapter provides an excellent overview of developments in the field of biodegradable plastic research.

Eriksen, Marcus. 2017. *Junk Raft: An Ocean Voyage and a Rising Tide of Activism to Fight Plastic Pollution.* Boston: Beacon Press.

This is a somewhat informal story of the author and his wife's trips across the oceans studying the presence and effects of plastic pollution. They later founded the 5Gyres Institute for the study of marine plastic pollution.

Franco-García, Maria-Laura, Jorge Carlos Carpio-Aguilar, and Hans Bressers, eds. 2019. *Towards Zero Waste: Circular Economy Boost, Waste to Resources*. Cham, Switzerland: Springer.

The essays in this book describe and discuss the implementation of zero waste and circular economy plans in countries around the world, such as China, Indonesia, Mexico, Netherlands, and Romania. It discusses the successes and challenges of such programs.

Gilbert, Marianne. 2017. *Brydson's Plastics Materials*. 8th ed. Oxford, UK; Cambridge, MA: Butterworth-Heinemann.

This highly regarded book provides a somewhat technical introduction to nearly every phase of plastic production, divided largely on the basis of various kinds of plastic.

Goel, Sudha, ed. 2017. *Advances in Solid and Hazardous Waste Management*. New Delhi: Springer.

This book provides more than a dozen essays on specific advances that have been made in the field of waste management technology. Some topics relate to the use of remote sensing and the GIS systems in waste management, leaching in pond ash, degradation of plastics, and special issues related to electronic wastes.

Goodship, Vannessa. 2007. *Introduction to Plastics Recycling*. Shawbury, UK: Smithers Rapra.

This book provides a good general introduction to the technologies of plastic recycling, along with some problems associated with the procedures, and an outlook to the future.

Gregory, Murray R., and Anthony L. Andrady. 2003. "Plastics in the Marine Environment." In Anthony L. Andrady, ed. *Plastics and the Marine Environment*. New York: John Wiley & Sons, Chapter 10.

This chapter is of special interest because it is apparently the first time the term *mesoplastic* was introduced to the literature. The definition suggested by the authors is a small piece of plastic greater than 5 millimeters in size, but smaller than a piece of macroplastic, defined as large enough to see well with the naked eye. The term is not widely used in the field. See, as one example, Isobe, Atsuhiko, et al. 2014. "Selective Transport of Microplastics and Mesoplastics by Drifting in Coastal Waters." *Marine Pollution Bulletin*. 89(1–2): 324–330. https://doi .org/10.1016/j.marpolbul.2014.09.041.

Guillet, James E. 1990/2018. "Photodegradable Plastics." In Sumner A. Barenberg, ed. *Degradable Materials: Perspectives, Issues, and Opportunities*. Boca Raton, FL: CRC Press, 55–58.
This somewhat dated chapter is an excellent, if somewhat technical, introduction to the topic of photodegradable plastics and their use in dealing with the growing problem of plastic wastes.

Harrison, R. M., and R. E. Hester. 2018. *Plastics and the Environment*. London: Royal Society of Chemistry.
This volume is of special interest because, in addition to the usual review of marine plastic pollution, it covers topics such as microplastics, nanoplastics, plasticizers, and plastic additives and their effects on human health.

Jaime, Cristina Garcia, ed. 2018. *Plastics in Packaging*. Oakville, ON, Canada: Delve Publishing.
Packaging is by far the most important application of plastics today. Yet, this widespread use of the material has become associated with serious social, economic, and environmental issues. The author explores the history of plastic packaging, as well as its current status, modes of production, associated problems, and possible solutions.

Karapanagioti, Hrissi K., and Ioannis K Kalavrouziotis, eds. 2019. *Microplastics in Water and Wastewater*. London: IWA Publishing.

The chapters in this book deal with topics such as microplastics in the human water cycle, wastewater treatment plants, and sludge; association of toxic compounds with microplastics; pollution of beaches and watercourses by microplastics; possible effects on plants from microplastics; and the need for a global microplastics strategy.

Koelmans, Albert A., Ellen Besseling, and Won J. Shim. "Nanoplastics in the Aquatic Environment. Critical Review." In Melanie Bergmann, Lars Gutow, and Michael Klages, eds. *Marine Anthropogenic Litter*. Cham, Switzerland: Springer, 325–340.

This chapter contains what is thought to be the first formal use of the term *nanoplastics* in the professional literature. The article itself is an excellent introduction to the topic with a review of the information currently available to researchers on nanoplastics.

Lackner, Maximilian. 2005/2015. "Bioplastics." In Raymond E. Kirk and Donald F. Othmer. *Kirk-Othmer Encyclopedia of Chemical Technology*. New York: Wiley-VCH.

This article provides a very nice general overview of the history, methodologies, and developments in the field of biodegradable plastics.

Lerner, Steve. 2012. *Sacrifice Zones: The Front Lines of Toxic Chemical Exposure in the United States*. Cambridge, MA: MIT Press.

The disproportionate harm caused to communities of color and poor people by industrial operations in their neighborhood is now well documented (although not necessarily dealt with very effectively). Chapter 5 of this book focuses on the issues posed by the presence of a plastics plant in such a neighborhood in Addyston, Ohio.

Leslie, H. A. 2014. *Review of Microplastics in Cosmetics*. Amsterdam, The Netherlands: Institute for Environmental Studies.

http://www.ivm.vu.nl/en/Images/Plastic_ingredients_in _Cosmetics_072014_FINAL_tcm234-409859.pdf.
This book focuses on three main topics; plastic ingredients in personal care and cosmetic products, functions of microplastics in these products, and environmental fate and effects of these microplastics.

Letcher, Trevor M. 2020. *Plastic Waste and Recycling: Environmental Impact, Societal Issues, Prevention, and Solutions.* Amsterdam, The Netherlands: Academic Press.
The articles in this book cover all aspects of plastic recycling from the specific challenges posed by each type of plastic to the steps involved in recycling, to challenges posed by plastic recycling in today's world.

Lundquist, Lars, et al. 2000. *Life Cycle Engineering of Plastics: Technology, Economy and Environment.* Kidlington, UK: Elsevier.
This book provides an excellent, if somewhat dated, general introduction to the concept of life cycle assessment and its applications to plastics.

Madden, Odile, et al. eds. 2017. *The Age of Plastic: Ingenuity and Responsibility: Proceedings of the 2012 MCI Symposium.* Washington, DC: Smithsonian Institution Scholarly Press. https://www.researchgate.net/publication/322212487_The _Age_of_Plastic_Ingenuity_and_Responsibility.
This collection of essays covers all conceivable aspects of the rise and current role of plastics in society, including topics such as advanced materials used in space suit design, plastics and the food culture, the role of the Ford company in the development of a plastic economy, preserving plastics for art objects, debris in the marine environment, and closing the loop with plastics.

McCallum, Will. 2018. *How to Give up Plastic: A Guide to Changing the World, One Plastic Bottle at a Time.* New York: Penguin Books.

This book is a very practical guide as to how it is possible to eliminate the use of plastic objects in every part of a person's life.

Mossman, S. T. I., and P. J. T. Morris. 1994. *The Development of Plastics.* Cambridge, UK: Royal Society of Chemistry.
This publication contains the proceedings of a symposium on the history of synthetic materials, held in conjunction with the Plastic Historical Society in April 1993. It contains articles on topics such as Victorian plastics; foundations of the plastic industry, Parkesine and Celluloid; Britain and the Bakelite Revolution; plastics and design in the United States, 1925–1935; the early history of polythene; and plastics and prosperity in the period from 1945 to1970.

Niaounakis, Michael. 2017. *Management of Marine Plastic Debris: Prevention, Recycling, and Waste Management.* Oxford, UK; Cambridge, MA: Elsevier.
This book provides an excellent overview of most issues relating to the management of plastic wastes in the oceans, with chapters on environmental, social, and economic impacts; degradation of plastics in the marine environment; prevention and mitigation; and regulatory framework.

Niceforo, Marina. 2019. *The Terminology of Marine Pollution by Plastics and Microplastics.* Newcastle upon Tyne, UK: Cambridge Scholars Publisher.
The subject of this book is somewhat esoteric, but it turns out to be a fascinating review of the history and current status of marine pollution caused by plastics and microplastics.

Nollet, Leo M. L., and Khwaja Salahuddin Siddiqi, eds. 2020. *Analysis of Nanoplastics and Microplastics in Food.* Boca Raton, FL: CRC Press.

This collection of articles deals with topics such as the origin, fate, and effects of nanoplastics and microplastics; their presence in food; methods of analysis; and reviews of the main types of plastics involved.

Orzolek, Michael, ed. 2017. *A Guide to the Manufacture, Performance, and Potential of Plastics in Agriculture.* Oxford, UK; Cambridge, MA: William Andrew.
This excellent reference book provides information about just every conceivable application of plastics in the agricultural industry, including plastic mulch, drip irrigation, tunnels, greenhouses, and row covers, along with chapters on the early history of the topic and good agricultural practices with plastics.

Plunkett, Jack W. 2020. *Plunkett's Chemicals, Coatings & Plastics Industry Almanac 2020: The Only Comprehensive Guide to the Chemicals Industry.* Houston, TX: Plunkett Research, Ltd.
While providing a good overall introduction to the basics of plastics research, this book focuses on business aspects of the topic. It includes statistics, business trends, new business models, and related topics in the plastics industry.

Rudolph, Natalie, Raphael Kiesel, and Chuanchom Aumnate. 2017. *Understanding Plastics Recycling Economic: Ecological, and Technical Aspects of Plastic Waste Handling.* Cincinnati, OH: Hanser Publications.
This text deals with virtually all aspects of the plastic waste issue, including sections on municipal solid wastes, recycling, economic factors and considerations, and environmental issues.

Shahnawaz, Mohd, Manisha K. Sangale, and Avinhash B. Ade. 2019. *Bioremediation Technology for Plastic Waste.* Singapore: Springer Verlag.

This excellent resource provides a thorough discussion of all aspects of plastic waste, including microplastics, plastic waste disposal, case studies of plastic waste degradation, the role of bacteria in bioremediation of plastics, and social awareness of plastic waste issues.

Subramanian, Muralisrinivasan Natamai. 2019. *Plastics Waste Management: Processing and Disposal.* 2nd ed. New York: John Wiley & Sons.

This book consists of 12 sections covering topics such as kinds of plastics, plastic additives, plastics and the environment, plastic processing technologies, plastic wastes, economy and the recycling market, life cycle assessment, and case studies.

Szaky, Tom, et al. 2019. *The Future of Packaging: From Linear to Circular.* Oakland, CA: Berrett-Koehler Publishers.

The 15 essays in this book focus on the problem of waste products from packaging materials. Some topics include "The State of the Recycling Industry," "Who Is Responsible for Recycling Packing?," "Designing Packages for the Simpler Recycler," "The Myth of Biodegradability," "The Forgotten Ones: Pre-Consumer Waste," and "Value for Business in the Circular Economy."

Thakur, Vijay Kumar, Manju Kumari Thakur, and Michael R. Kessler, eds. 2017. *Soy-based Bioplastics.* Shewbury, UK: Smithers Rapra Technology Ltd.

This book provides a very detailed and technical discussion of research on the use of soy products in the manufacture of new biodegradable plastics.

Wagner, Martin, and Scott Lambert, eds. 2018. *Freshwater Microplastics: Emerging Environmental Contaminants?* Cham, Switzerland: Springer. https://www.researchgate.net /publication/321225506_Freshwater_Microplastics_Emerging _Environmental_Contaminants.

The study of freshwater plastic pollution is probably somewhat less popular than is the study of comparable marine pollution. The articles in this book are 13 topics of special relevance to this issue, including sources and fates of microplastics in urban environments and in inland water in Asia; interactions of microplastics and freshwater biota; microplastic-associated biofilms; and issues of regulation and management of microplastic pollution.

Wani, Khursheed Ahmad, Lutfah Ariana, and S. M. Zuber, eds. 2020. *Handbook of Research on Environmental and Human Health Impacts of Plastic Pollution*. Hershey, PA: USA Engineering Science Reference.

The value of this book lies in its coverage of virtually every basic issue in plastics pollution, including its effects on the environment and human health, impact on marine organisms, special issues of medical wastes, analysis of the role of specific compounds such as bisphenol A and phthalates, and strategies for waste plastic management.

## Articles

The topic of plastics and microplastics is the main focus of several journals. Among those journals are the following:

*Environmental Science & Technology*: ISSN (print): 0013-936X; ISSN (online): 1520-5851

*International Journal of Environmental Research and Public Health*: ISSN (print): 1661-7827; ISSN (online): 1660-4601

*International Journal of Plastic and Polymer Technology*: ISSN (print): 2249-801X

*International Journal of Plastics Technology*: ISSN (print): 0972-656X; ISSN (online): 0975-072X

*Journal of Elastomers & Plastics*: ISSN (print): 0095-2443; ISSN (online): 1530-8006

*Plastics Engineering*: ISSN (online): 1941-9635

*Plastics News*: https://www.plasticsnews.com/

*Plastics, Rubber and Composites: Macromolecular Engineering*: ISSN (print): 1465-8011

*Plastics Today*: https://www.plasticstoday.com/

*Progress in Polymer Science*: ISSN (print): 0079-6700

*Science of the Total Environment*: ISSN (online): 0048-9697

Andrady, Anthony L. 2011. "Microplastics in the Marine Environment." *Marine Pollution Bulletin.* 62(8): 1596–1605. https://doi.org/10.1016/j.marpolbul.2011.05.030.

> This slightly dated article provides an excellent overview of the problems of microplastics and nanoplastics in the oceans, with sections on occurrence, composition, toxicity, and degradation under ocean conditions.

Bonanno, Giuseppe, and Martina Orlando-Bonaca. 2018. "Ten Inconvenient Questions about Plastics in the Sea." *Environmental Science & Policy.* 85: 146–154.

> This article might also be titled "Some Important Things We Don't Know about Plastics in the Oceans." It raises and then discusses questions such as how we gain realistic estimates of the problem, what we know about the sources of plastic pollution in the seas, which categories of plastics are most likely to reach the oceans, and what is the fate of plastics in the seas.

Borrelle, Stephanie B., et al. 2017. "Opinion: Why We Need an International Agreement on Marine Plastic Pollution." *PNAS.* 114(38): 9994–9997; https://doi.org/10.1073/pnas.1714450114.

> This article focuses on the need for international agreement on methods for reducing plastic pollution in the

oceans. But it is also a good introduction to the scope and nature of that problem.

Brandl, Helmut, and Petra Püchner. 1991. "Biodegradation of Plastic Bottles Made from 'Biopol' in an Aquatic Ecosystem under in Situ Conditions." *Biodegradation*. 2: 237–243.
This article describes research on the ability of a new plastic, called Biopol, to degrade under normal conditions in Lake Lugano, Switzerland.

Budsaereechai, Supattra, Andrew J. Hunt, and Yuvarat Ngernyen. 2019. "Catalytic Pyrolysis of Plastic Waste for the Production of Liquid Fuels for Engines." *RSC Advances*. 9: 5488. https://pubs.rsc.org/en/content/articlepdf/2019/ra/c8ra10058f.
There has been considerable interest in the pyrolysis of plastics as a way of dealing with the problem of plastic wastes, along with their subsequent conversion of products that can be used for commercial fuels. This article summarizes some of the most recent research in that field.

Carpenter, Edward J., et al. 1972. "Polystyrene Spherules in Coastal Waters." *Science*. 178(4062): 749–750. https://doi:10.1126/science.178.4062.749.
The authors of this report provide what may be the first mention of microplastics in ocean waters.

Chatterjee, Subhankar, and Shivika Sharma. 2019. "Microplastics in Our Oceans and Marine Health." *Field Actions Science Report*. Special Issue 19: 54–61. https://journals.openedition.org/factsreports/5257.
The authors provide a good, general introduction to problems associated with the presence of microplastic pollution in the oceans.

Chen, Qiqing, et al. 2018. "Pollutants in Plastics within the North Pacific Subtropical Gyre." *Environmental Science and Technology*. 52(2): 446–456. https://doi.org/10.1021/acs.est.7b04682.

The authors of this article report on their studies of plastics in one region of the Great Pacific Garbage Patch. They studied the composition of plastics in this region, potential toxic effects on marine organisms there, and possible issues of bioaccumulation in the patch.

Daglen, Bevin C., and David R. Tyler. 2009. "Photodegradable Plastics: End-of-life Design Principles." *Green Chemistry Letters and Reviews.* 3(2): 69–82. https://doi.org/10.1080/17518250903506723.
    This article provides a good introduction to the concept of photodegradable plastics, their desired properties, and the methods by which they can be produced.

Farady, Susan E. 2019. "Microplastics as a New, Ubiquitous Pollutant: Strategies to Anticipate Management and Advise Seafood Consumers." *Marine Policy.* 104: 103–107.
    Research indicates that microplastics are now being found in aquatic organisms consumed by humans as foods. The author explores the health issues posed by this discovery and suggests some actions that can be taken to protect consumers from the deleterious effects of the practice.

Fatima, Zernab, and Roohi. 2019. "Smart Approach of Solid Waste Management for Recycling of Polymers: A Review." *Current Biochemical Engineering.* 5(1): 4-11.
    The authors suggest a relatively new approach to the recycling of plastics, namely the use of macroplastics in the manufacture of concrete and wood plastic composites.

Fellet, Melissae. 2018. "Improving an Enzyme for Better Plastic Recycling." *C&EN.* 96(10): 7–7.
    This article provides a modestly challenging technical explanation of the way enzymes can be designed to be used in the manufacture of biodegradable plastics.

Ferreira, Inês, et al. 2019. "Nanoplastics and Marine Organisms: What Has Been Studied?" *Environmental Toxicology and Pharmacology.* 67: 1–7. https://doi.org/10.1016/j.etap.2019.01.006.

> The authors point out that little or no information is available on the effects of nanoplastics on marine vertebrates (fish), but that more research has been done on other marine species. They summarize and comment on that research in this paper.

Filho, Walter Leal, et al. 2019. "Plastic Debris on Pacific Islands: Ecological and Health Implications." *Science of the Total Environment.* 670: 181–187.

> Plastic pollution of the ocean has been the subject of a considerable number of research works. The effect of this phenomenon on the beaches and land area of islands located in the oceans is less studied. This study finds that such effects are significant and need to be addressed.

Foreinis, Spyros. 2020. "How Small Daily Choices Play a Huge Role in Climate Change: The Disposable Paper Cup Environmental Bane." *Journal of Cleaner Production.* 255: 120294. https://doi.org/10.1016/j.jclepro.2020.120294.

> Critics sometimes suggest that bans on plastic bags, straws, cups, or other items are a waste of time, representing only a "drop in the bucket" in dealing with the overall problem of plastic wastes. This article shows the error in that line of thinking with a discussion of the effects of no longer using paper cups, a savings that presumably would be equivalent to that for plastic cups.

Gall, S. C., and R. C. Thompson. 2015. "The Impact of Debris on Marine Life." *Marine Pollution.* 92(1–2): 170–179. https://doi.org/10.1016/j.marpolbul.2014.12.041.

> This paper provides "[a]n extensive literature search [that] reviewed the current state of knowledge on the effects of

marine debris on marine organisms" as found in 340 original publications on the topic.

Gigault, Julien, et al. 2018. "Current Opinion: What Is a Nanoplastic?" *Environmental Pollution*. 235: 1030–1034. https://doi.org/10.1016/j.envpol.2018.01.024.
    The authors take note of the fact that there is currently no widely agreed-upon definition for nanoparticles, a problem that they attempt to resolve in this article. They also note that research on the occurrence and effects of nanoplastics is sparse and suggest possible research topics in the field.

Gove, Jamison M., et al. 2019. "Prey-Size Plastics Are Invading Larval Fish Nurseries." *PNAS*. 116(48): 24143–24149. https://doi.org/10.1073/pnas.1907496116.
    This article reports on the discovery of microplastic particles the size of prey for fish in the larval stage. It discusses the impact on this form of pollution on developing fish stocks.

Horton, Alice A., et al. 2017. "Microplastics in Freshwater and Terrestrial Environments: Evaluating the Current Understanding to Identify the Knowledge Gaps and Future Research Priorities." *Science of the Total Environment*. 586: 127–141. https://www.researchgate.net/publication/313358945_Microplastics_in_freshwater_and_terrestrial_environments_Evaluating_the_current_understanding_to_identify_the_knowledge_gaps_and_future_research_priorities.
    Plastic and microplastic pollution is a relatively well-studied problem. Much less is known about the pollution of lakes, streams, and other freshwater bodies. This article summarizes the knowledge available so far on that topic and the areas of research that should be considered for the future.

Huysman, Sofie, et al. 2015. "The Recyclability Benefit Rate of Closed-Loop and Open-Loop Systems: A Case Study on

Plastic Recycling in Flanders." *Resources, Conservation and Recycling.* 101: 53–60. https://doi.org/10.1016/j.resconrec .2015.05.014.

> This widely quoted paper provides quantitative data on the relative efficiency of four types of plastic treatment systems: incineration, landfill, open-loop, and closed-loop systems. The latter two are found to be more efficient modes of operation than the former two.

Ivar do Sul, Juliana A., and Monica F. Costa. 2014. "The Present and Future of Microplastic Pollution in the Marine Environment." *Environmental Pollution.* 185: 352–364. https://doi .org/10.1016/j.envpol.2013.10.036.

> The presence and effects of microplastics in the marine environment have become a very active field of research in the marine sciences. This article summarizes the main themes and findings of that research as of 2014.

Jambeck, Jenna R., et al. 2015. "Plastic Waste Inputs from Land into the Ocean." *Science.* 347(6223): 768–770. https:// wedocs.unep.org/bitstream/handle/20.500.11822/17969 /Plastic_waste_inputs_from_land_into_the_ocean.pdf.

> This research team attempts to estimate the amount of plastic waste released into the oceans from land-based sources. They conclude that the problem of plastic waste could increase by as much as 10 times by 2025.

Kampmann, Marie, et al. 2019. "Quality Assessment and Circularity Potential of Recovery Systems for Household Plastic Waste." *Journal of Industrial Ecology.* 23(1): 156–168. https:// onlinelibrary.wiley.com/doi/pdf/10.1111/jiec.12822.

> Individuals and households are often targeted as major sites at which better plastic waste management procedures can be adopted. But many technical and practical problems are associated with the development of successful household plastic waste programs. This article reviews 84

different recovery scenarios that can be used or can contribute to the solution of this problem.

Kandziora, J. H., et al. 2019. "The Important Role of Marine Debris Networks to Prevent and Reduce Ocean." *Marine Pollution Bulletin.* 141: 657–662.

Beginning in the second decade of the twenty-first century, several nations and regions began to appreciate the need for regional systems for the monitoring of ocean wastes. In 2018, at the Sixth International Marine Debris Conference, held in San Diego, a group of regional programs decided to band together to form a single, worldwide system for carrying out this objective. The program, called the International Waste Platform (IWP), was created at this meeting. The structure, goals, and activities of the IWP are described in this article.

Karan, Hakan, et al. 2019. "Green Bioplastics as Part of a Circular Bioeconomy." *Trends in Plant Science.* 24(3): 237–249. https://doi.org/10.1016/j.tplants.2018.11.010.

This paper deals with the types of bioplastics currently being made, their chemical synthesis, progress in scaling up laboratory procedures, properties and uses of bioplastics, current regulatory framework of bioplastics, and key barriers and opportunities in the field.

Koči, Vladimir. 2019. "Comparisons of Environmental Impacts between Wood and Plastic Transport Pallets." *Science of the Total Environment.* 686: 514–528. https://doi.org/10.1016/j .scitotenv.2019.05.472.

One of the most common suggestions for reducing the amount of plastic waste as part of a circular economy is reverting to traditional materials that have been used in the past. This review compares the relative value of wood and virgin and secondary plastic in the manufacture and

use of transport pallets. The author finds that wood pallets are superior to both types of plastic pallets.

Kore, Sudarshan D. 2019. "Sustainable Utilization of Plastic
Waste in Concrete Mixes—A Review." *Journal of Building Materials and Structures*. 5(2): 212–217. https://www.researchgate
.net/publication/330555546_Sustainable_Utilization_of
_Plastic_Waste_in_Concrete_Mixes-a_Review.
    The conversion of plastic wastes to useful products has
    been a particularly difficult challenge for researchers. This
    article summarizes some of the suggestions that have been
    made for using waste plastics in the formation of concrete
    materials.

Laist, David. 1995. "Marine Debris Entanglement and Ghost
Fishing: A Cryptic and Significant Type of Bycatch?" Conference Publication: Solving Bycatch: Considerations for Today
and Tomorrow. Research Gate. https://www.researchgate.net
/publication/259929224_Marine_Debris_Entanglement
_and_Ghost_Fishing_A_Cryptic_and_Significant_Type_of
_Bycatch.
    This somewhat dated paper reviews the data for and evi-
    dence about entanglement, especially with ghost fishing
    gear, and its effects on marine organisms.

Lebreton, L., et al. 2018. "Evidence That the Great Pacific
Garbage Patch Is Rapidly Accumulating Plastic." *Scientific Reports*. 8(1): 1–15. https://www.nature.com/articles
/s41598-018-22939-w.pdf.
    This report focuses on studies of the source, amount, and
    character of plastic waste pollution in the Pacific Ocean.
    It deals primarily with the region generally known as the
    Great Pacific Garbage Patch. The research team reports
    that the amount of debris in the region is 4–16 times as
    great as previously reported.

Lebreton, Laurent, and Anthony Andrady. 2019. "Future Scenarios of Global Plastic Waste Generation and Disposal." *Palgrave Communications.* 5(1): 1–11. https://www.researchgate .net/publication/330542951_Future_scenarios_of_global _plastic_waste_generation_and_disposal.

> Using high-resolution observational technology, the authors project the status of plastic pollution in the oceans in 2060. They conclude that developing nations are at greatest risk of harmful social and economic effects by their projections. They suggest some ways by which worst-case scenarios can be avoided in such regions.

Li, Jingyi, Huihui Liu, and J. Paul Chen. 2018. "Microplastics in Freshwater Systems: A Review on Occurrence, Environmental Effects, and Methods for Microplastics Detection." *Water Research.* 137: 362–374. https://doi.org/10.1016/j.watres.2017 .12.056.

> This article provides a general overview of current information on the effects of microplastics on fish.

Maximenko, Nikolai, et al. 2019. "Toward the Integrated Marine Debris Observing System." *Frontiers in Marine Science.* 6. https://doi.org/10.3389/fmars.2019.00447.

> This article describes a projected new program for charting the quantity and location of wastes present in the oceans. It reviews the properties and distribution of wastes and the technologies that will be used in monitoring their location and movement through the seas.

Meikle, Jeffrey L. 1992. "Into the Fourth Kingdom: Representations of Plastic Materials, 1920–1950." *Journal of Design History.* 5(3): 173–182. https://www.jstor.org/stable/1315836.

> The author argues that there has been a "revolutionary" feature to the plastics industry ever since its origins in the 1920s. He pursues the social, political, and philosophical

effects of many of the new manifestations of plastic technology on society.

Moharir, Rucha V., and Sunil Kumar. 2019. "Challenges Associated with Plastic Waste Disposal and Allied Microbial Routes for Its Effective Degradation: a Comprehensive Review." *Journal of Cleaner Production*. 208: 65–76.

The authors review the status of the world's plastic waste problems and the effectiveness of existing technologies to deal with this problem. They conclude that other approaches are needed, and they review the methodology and examples of microbial degradation for dealing with the problem.

Monteiro, Raqueline C. P., Juliana A. Ivar do Sul, and Monica F. Costa. 2018. "Plastic Pollution in Islands of the Atlantic Ocean." *Environmental Pollution*. 238: 103–110.

Studies on the presence and impact of plastics on the oceans are increasing in popularity. Somewhat less attention is being paid to effects on oceanic islands. This article provides a review of the most recent and relevant studies on the topic.

Napper, Imogen E., and Richard C. Thompson. 2019. "Environmental Deterioration of Biodegradable, Oxo-biodegradable, Compostable, and Conventional Plastic Carrier Bags in the Sea, Soil, and Open-Air Over a 3-Year Period." *Environmental Science & Technology*. 53(9): 4775–4783.

The author explores the fate of various types of biodegradable plastics under normal environmental conditions over a three-year period. They find quite different results for the various types of plastics.

Okan, Meltem, et al. 2019. "Current Approaches to Waste Polymer Utilization and Minimization: A Review." *Journal of*

*Chemical Technology & Biotechnology.* 94(1): 8–21. https://onlinelibrary.wiley.com/doi/epdf/10.1002/jctb.5778.
> Researchers are exploring a variety of methods for treating plastic wastes in an environmentally sensitive and sustainable way. This paper reviews those methods, with special attention to a few methods that are entirely new in their approach to the problem.

Paxton, Naomi C., et al. 2019. "Biomedical Applications of Polyethylene." *European Polymer Journal.* 118: 412–428. https://doi.org/10.1016/j.eurpolymj.2019.05.037.
> This article provides a nice review of the wide applications of polyethylene in the field of medicine.

Ragaert, Kim, Laurens Delva, and KevinVan Geem. 2017. "Mechanical and Chemical Recycling of Solid Plastic Waste." *Waste Management.* 69: 24–58.
> Mechanical and chemical recycling are the two major methods for dealing with plastic wastes. This article discusses the methodologies involved with each, along with a general overview of the ways in which such wastes are dealt with by waste management programs.

Rahmeem, Ademola Bolanle, et al. 2019. "Current Developments in Chemical Recycling of Post-consumer Polyethylene Terephthalate Wastes for New Materials Production: A Review." *Journal of Cleaner Production.* 225: 1052–1064. https://doi.org/10.1016/j.jclepro.2019.04.019.
> Chemical recycling poses some of the most difficult challenges in plastic waste recycling among all methods. This article reviews the current status of affairs in that field.

Rajendran, Saravanakumar, et al. 2015. "Programmed Photodegradation of Polymeric/Oligomeric Materials Derived from Renewable Bioresources." *Angewandte Chemie International Edition.* 54(4): 1159–1163.

The authors describe a method by which photodegradable plastics can be produced and the results of their research on these products when exposed to environmental conditions.

Ramirez, Abel, and Babu George. 2918. "Plastic Recycling and Waste Reduction in the Hospitality Industry." *Economics, Management and Sustainability.* 4(1): 6–20. https://doi.org/10.14254/jems.2019.4-1.1.

Restaurants, hotels, and other hospitality businesses are a major contributor to the problem of plastic wastes in today's world. This article discusses some of those problems and actions that are being taken to deal with plastic wastes within the industry.

Reade, Lou. 2018. "Plastics All at Sea." *SCI.* https://www.soci.org/chemistry-and-industry/cni-data/2018/2/plastics-all-at-sea.

This article begins with a review of research that suggests that levels of microplastics have not changed over the past 30 years. It then goes on to discuss the problem in general, along with some interesting proposals that have been made for solving the problem.

Rhodes, Christopher J. 2019. "Solving the Plastic Problem: from Cradle to Grave, to Reincarnation." *Science Progress.* 102(3): 218–248. https://journals.sagepub.com/doi/pdf/10.1177/0036850419867204.

This article provides a wide-ranging discussion of the scope of the plastic pollution problem methods that are used and have been suggested for dealing with that problem, the relative success of each approach, the role of plastics pollution in climate change, and other issues.

Ryan, Peter J. 2015. "A Brief History of Marine Litter Research." In Melanie Bergmann, Lars Gutow, and Michael

Klages, eds. 2015. *Marine Anthropogenic Litter.* Cham, Switzerland: Springer International Publishing, 1–25. https://link.springer.com/chapter/10.1007/978-3-319-16510-3_1.

This article is a superb summary of the research conducted over the years on the presence, character, amount, and effects of litter on the marine environment. This is required reading for anyone interested in the topic.

Shim, Won Joon, and Richard C. Thompson. 2015. "Microplastics in the Ocean." *Archives of Environmental Contamination and Toxicology.* 69(3): full.

This special issue of the journal includes papers on several very specific topics in the field, such as the ingestion of microplastics by zooplankton, characteristics of plastic marine debris on the beaches of South Korea, and the qualitative analysis of additives in marine microplastics.

Singh, Narinder, et al. 2017. "Recycling of Plastic Solid Waste: A State of Art Review and Future Applications." *Composites Part B: Engineering.* 115: 409–422. https://doi.org/10.1016/j.compositesb.2016.09.013.

The challenges faced by accumulating amounts of plastic waste have encouraged researchers to explore a variety of ways with which to deal with this problem. This paper summarizes the work being done in the field.

Stafford, Richard, and Peter J. S. Jones. 2019. "Viewpoint—Ocean Plastic Pollution: A Convenient but Distracting Truth?" *Marine Policy.* 103: 187–191. https://doi.org/10.1016/j.marpol.2019.02.003.

The authors argue that plastic pollution of the oceans is an important problem, but not of the scale of climate change and overfishing. They suggest that nations and corporations should choose to focus on plastics pollution as a way of proving their concern about "green" issues. But they miss some of the most important actions needed for ocean health in the process. Also, see response to this article at Avery-Gomm, Stephanie, et al. 2019. "There Is Nothing Convenient about Plastic Pollution. Rejoinder to Stafford and Jones 'Viewpoint—Ocean Plastic Pollution: A Convenient but Distracting Truth?'" *Marine Policy*. 106: https://doi.org/10.1016/j.marpol.2019.103552.

Steinmetz, Zacharias, et al. 2016. "Plastic Mulching in Agriculture. Trading Short-term Agronomic Benefits for Long-term Soil Degradation?" *Science of the Total Environment*. 550: 690–705. https://doi.org/10.1016/j.scitotenv.2016.01.153.
Mulching with plastic sheets and other materials has become a common and useful agricultural technology. But loss of plastic particles and microplastics to the surrounding ground have led to serious problems for the soil and organisms that live there. This article reviews the current knowledge about this problem, along with an assessment of the risks and benefits of plastic mulching in agricultural operations.

Thompson, Richard C., et al. 2004. "Lost at Sea: Where Is All the Plastic?" *Science*. 304(5672): 838. https://doi.org/10.1126/science.1094559.
Thompson is often credited with having first used the term *microplastics* in this article. It describes some of the earliest relatively straightforward research on the occurrence and nature of microplastics.

Wang, Jiao, et al. 2019. "Microplastics as Contaminants in the Soil Environment: A Mini-review." *Science of the Total Environment*. 691: 848–857.
The authors provide a general overview of the kind and amount of microplastics found in the soil, the main sources of those microplastics, and the possible effects on the soil and organisms associated with the soil, along with possible threats to human health.

Wang, Wanli, et al. 2019. "Current Influence of China's Ban on Plastic Waste Imports." *Waste Disposal & Sustainable Energy*. 1(1): 67–78. https://link.springer.com/article/10.1007 /s42768-019-00005-z.

For many years, the United States, European Union, and other developed countries have relied on China's acceptance and recycling of plastic wastes for dealing with a major waste management problem. The authors here review actions being taken and planned by those nations to adjust to China's recent decision to withdraw from the plastic recycling market.

Xanthose, Dirk, and Tony R. Walker. 2017. "International Policies to Reduce Plastic Marine Pollution from Single-use Plastics (Plastic Bags and Microbeads): A Review." *Marine Pollution Bulletin*. 118(1–2): 17–26. https://doi.org/10.1016/j .marpolbul.2017.02.048.

Governmental bodies at the national, state, and local levels have devised a variety of legal methods for dealing with the surge of plastic wastes worldwide. This article summarizes information about these programs.

The Yoyo Team. 2019. "Yoyo: Recycling All Plastic. Impossible? We've Already Started!" *Field Actions Science Report*. 19: 92–95. https://journals.openedition.org/factsreports/5450#tocto2n1.

This French organization was created on the basis of a philosophy that individuals can make significant contributions to the problems of solving plastic pollution. In this article, they outline their own history, their approach to plastic recycling, and results of their work thus far.

Zheng, Jiajia, and Sangwon Suh. 2019. "Strategies to Reduce the Global Carbon Footprint of Plastics." *Nature Climate Change*. 9(5): 374–378. https://www.polybags.co.uk/environmentally-friendly /strategies-to-reduce-the-global-carbon-footprint-of-plastics.pdf.

This article is an excellent example of the use of life cycle assessment in comparing two or more methodologies for the use of plastic materials. For definitions of life cycle assessment, see Defining Life Cycle Assessment n.d.

Zhu, Bo-Kai, et al. 2018. "Exposure to Nanoplastics Disturbs the Gut Microbiome in the Soil Oligochaete *Enchytraeus Crypticus*." *Environmental Pollution*. 239: 408–415. https://doi.org /10.1016/j.envpol.2018.04.017.

The authors note that relatively little research has been conducted on the effects of microplastics and nanoplastics on terrestrial organisms. In this study, they determine the effects of nanoparticles of differing sizes on the biological function of the *Enchytraeus crypticus* microbiome. They find changes in both bacterial diversity and structure of the organisms, as well as altered weight and reproduction patterns of the organisms.

## Reports

"Annex I. UNEA Resolution 1/6 Marine Plastic Debris and Microplastics." 2014. First UN Environment Assembly of the UN Environment Programme. https://wedocs.unep.org /bitstream/handle/20.500.11822/17562/Marine%20Plastic %20Debris%20and%20Microplastic%20Technical %20Report%20Advance%20Copy%20Ann.pdf?

This document presents a detailed and comprehensive review of essentially all existing data on the occurrence and distribution of plastics and microplastics in the world's oceans and some freshwater bodies, along with a summary of their effects on all of the studied marine organisms. For a background on the report, also see "Summary of the First UN Environment Assembly of the UN Environment Programme." 2014. IISD Reporting Services. https://enb.iisd.org/vol16/enb16122e.html.

Beaman, Joe, et al. 2016. "State of the Science White Paper: A Summary of Literature on the Chemical Toxicity of Plastics Pollution to Aquatic Life and Aquatic-Dependent Wildlife." U.S. Environmental Protection Agency. https://www.epa.gov /sites/production/files/2017-02/documents/tfw-trash_free _waters_plastics-aquatic-life-report-2016-12.pdf.

This report begins with a survey of the physical, chemical, and toxicological properties of various types of plastics and a summary of methods by which microplastics are transmitted to lakes and the oceans. It concludes with a review of the literature on the toxicological effects of microplastics on various types of organisms.

Boucher, Julien, and Damien Friot. 2017. "Primary Microplastics in the Oceans: A Global Evaluation of Sources." International Union for the Conservation of Nature. https://storyofstuff.org/wp-content/uploads/2017/02 /IUCN-report-Primary-microplastics-in-the-oceans.pdf.

This report provides a good general overview of current information available on the sources of microplastics. The appendices provide especially useful data on the topic.

"Discarded: Communities on the Frontlines of the Global Plastic Crisis." 2019. GAIA. https://wastetradestories.org /wp-content/uploads/2019/04/Discarded-Report-April-22.pdf.

For all the discussions on the problems posed by plastic waste, relatively little attention is paid to the differential effects of the problem of communities of differing economic status. This report explores the disproportionate effect of plastic waste on communities of color, poor people, and other disadvantaged groups in China, Malaysia, Thailand, and Indonesia.

"Ghost Gear: The Abandoned Fishing Nets Haunting Our Oceans." 2019. Greenpeace Germany. https://storage .googleapis.com/planet4-new-zealand-stateless/2019/11

/b97726c9-ghost_fishing_gear_report_en_single-page _051119.pdf.

This excellent report explains the types of fishing gear that represent a threat to ocean animals, the effects produced on the animals by that gear, and the international regulations dealing with the problem.

Kershaw, Peter John, et al. 2015. "Biodegradable Plastics & Marine Litter: Misconceptions, Concerns and Impacts on Marine Environments." United Nations Environment Programme. https://wedocs.unep.org/handle/20.500.11822 /7468.

This report examines the problem of plastic pollution of the oceans and the contribution that biodegradable plastics can make to solving that problem. The author concludes that "the adoption of plastic products labelled as 'biodegradable' will not bring about a significant decrease either in the quantity of plastic entering the ocean or the risk of physical and chemical impacts on the marine environment, on the balance of current scientific evidence."

Kershaw, Peter J., et al. 2016. "Marine Plastic Debris and Microplastics: Global Lessons and Research to Inspire Action and Guide Policy Change." Nairobi, Kenya: United Nations Environment Programme. https://searchworks.stanford.edu /view/12090852.

In 2014, the United Nations Environment Assembly recommended a study of worldwide issues associated with the presence of plastics and microplastics in the world's oceans. This report summarizes the findings of the committee appointed to conduct that study, along with a group of more than 50 recommendations for actions dealing with the problem.

"Legal Limits on Single-Use Plastics and Microplastics: A Global Review of National Laws and Regulations." 2018.

United Nations Environment Programme. https://wedocs .unep.org/bitstream/handle/20.500.11822/27113/plastics _limits.pdf.

This very detailed report summarizes laws and regulations involving paper bags, other single-use plastic items, and microbeads in all countries around the world.

Locock, Katherine E. S., et al. *The Recycled Plastics Market: Global Analysis and Trends*. Australia: CSIRO (Commonwealth Scientific and Industrial Research Organization).

This study attempts to outline the general features of the world's ocean plastic pollution problem with special emphasis on the economic factors involved in finding ways of dealing with that problem.

Lusher, Amy, Peter Hollman, and Jeremy Mendoza-Hill. 2017. "Microplastics in Fisheries and Aquaculture." Food and Agriculture Organization of the United Nations. http://www.fao .org/3/a-i7677e.pdf.

This report focuses on the problems posed by microplastics for aquatic organisms and food safety.

"Marine Debris: Understanding, Preventing and Mitigating the Significant Adverse Impacts on Marine and Coastal Biodiversity." 2016. Technical Series No.83. Secretariat of the Convention on Biological Diversity. https://www.cbd.int/doc /publications/cbd-ts-83-en.pdf.

This report provides a review of the most recent data available about the types of debris affecting the marine environment, the effects of these materials on marine organisms, and the steps that have been and can be taken to ameliorate this problem.

McIntosh, Naomi, et al. 2000. "Proceedings of the International Marine Debris Conference on Derelict Fishing Gear and the Ocean Environment." NOAA. National Marine Sanctuaries.

https://nmshawaiihumpbackwhale.blob.core.windows.net
/hawaiihumpbackwhale-prod/media/archive/documents/pdfs
_conferences/proceedings.pdf.
    This report provides a summary of the so-called Second International Marine Debris Conference, held in Honolulu, in 2000. It is a follow-up on the first such conference, held in 1984. See under Shomura and Yoshida.

"Microplastics in Drinking Water." 2019. World Health Organization. https://apps.who.int/iris/bitstream/handle/10665/326499/9789241516198-eng.pdf.
    This report provides a comprehensive review of the occurrence of microplastics in drinking water, along with an extended discussion of current evidence on possible and probable health effects of microplastics on human health.

"Microplastics in the Great Lakes: Workshop Report." 2016. International Joint Commission. https://legacyfiles.ijc.org/tinymce/uploaded/Microplastics_in_the_Great_Lakes_Workshop_Report_FINAL_September14-2016.pdf.
    This document is the report of a binational (Canada and the United States) commission created to study issues of microplastics in the Great Lakes. The commission came up with a set of 10 recommendations for better understanding and dealing with the problem of microplastic pollution in the Great Lakes.

Murphy, Margaret, et al. 2017. "Microplastics Expert Workshop Report." EPA Office of Wetlands, Oceans and Watersheds. https://www.epa.gov/sites/production/files/2018-03/documents/microplastics_expert_workshop_report_final_12-4-17.pdf.
    This report is a summary of a meeting held on June 28–29, 2017, to discuss research needs for a better understanding of the nature of the microplastics problem in U.S. waters and steps that should be taken to deal with that problem.

"A Sea of Problems: Impacts of Plastic Pollution on Oceans and Wildlife." 2020. Subcommittee on Water, Oceans, and Wildlife of the Committee on Natural Resources. U.S. House of Representatives. One Hundred Sixteenth Congress, first session, October 29, 2019. Washington, DC: U.S. Government Publishing Office. https://docs.house.gov/Committee/Calendar/ByEvent.aspx?EventID=110151.

    This document is a record of one of the most recent hearings in the U.S. Congress about the problems of plastic pollution in the oceans. It contains verbal and written evidence provided by witnesses appearing at the hearing.

Shomura, Richard S., and Howard O. Yoshida, eds. 1985. "Proceedngs of the Workshop on the Fate and Impact of Marine Debris 27–29 November 1984, Honolulu, Hawaii." NOAA Technical Memorandum NMFS. https://www.st.nmfs.noaa.gov/tm/swfc/swfc054.pdf.

    This report contains a summary of papers presented at an early conference on marine debris, the first of what was to become known as the International Conference on Marine Debris meetings.

"Summary of Expert Discussion Forum on Possible Human Health Risks from Microplastics in the Marine Environment." 2015. Marine Pollution Control Branch. Office of Wetlands, Oceans and Watersheds. U.S. Environmental Protection Agency. https://www.epa.gov/sites/production/files/2017-02/documents/tfw-microplastics_expert_forum_meeting_summary_2015-02-06.pdf.

    This report summarizes the findings of an EPA forum conducted on April 23, 2014, about possible health effects of microplastics on humans.

Tickell, Oliver. 2018. "International Law and Marine Plastic Pollution—Holding Offenders Accountable." Ocean Plastic Legal Initiative. http://www.apeuk.org/opli/.

This report emphasizes the serious nature of the world's ocean plastic pollution problem and recommends actions that can be taken to deal with that problem. Fundamentally, however, the report points out how existing treaties and agreements provide all the legal tools necessary to deal with the problem.

## Internet

Arney, Kat. 2017. "Casein." Chemistry World. https://www.chemistryworld.com/podcasts/casein/3007625.article.
This podcast tells the fascinating story of the discovery of casein plastics by Bavarian chemist Adolf Spitteler in 1897. (It was actually a cat that made the discovery!)

Backhaus, Thomas, and Martin Wagner. 2019. "Microplastics in the Environment: Much Ado about Nothing? A Debate." Global Challenges. https://doi.org/10.1002/gch2.201900022.
The two authors of this article began by debating on Twitter about the relative importance of microplastics in the environment. That debate later evolved into this journal article.

Baker, Maverick. 2018. "How to Eliminate Plastic Waste and Plastic Pollution with Science and Engineering." Interesting Engineering. https://interestingengineering.com/how-to-eliminate-plastic-waste-and-plastic-pollution-with-science-and-engineering.
Finding ways to convert waste plastics into useful new materials and objects is a major goal of much research today. This web page discusses about a dozen of the most promising ideas that have been developed thus far.

Benson, Imogen. 2019. "Research Reveals Damaging Impact of Microplastics on Human Health." Resource. https://resource.co/article/research-reveals-damaging-impact-microplastics-human-health.

This article reports on research by Dutch investigators on the effects of microplastics on the human body. One finding is that immune cells are destroyed by microplastics, suggesting the possibility of serious health consequences when exposed to such materials. (The Dutch article has not yet appeared in print as of this review.)

Boyd, Jane E. 2011. "Celluloid: The Eternal Substitute." Science History Institute. https://www.sciencehistory.org /distillations/celluloid-the-eternal-substitute.

This enjoyable essay discusses the origins of what the author calls "the first plastic," along with the changes it brought about in the culture of the time and since then.

Bruggers, James. 2019. "Plastics: The New Coal in Appalachia?" Inside Climate News. https://insideclimatenews.org /news/25022019/plastics-hub-appalachian-fracking-ethane -cracker-climate-change-health-ohio-river.

As demand for fossil fuels begins to decrease, energy companies are exploring new areas for their products. One such area is the production of plastics, which makes use of many of the same compounds produced during distillation and fracking operations. This article explores some of the developments now taking place in the Appalachian region to meet this objective.

Cao, Dongdong, et al. 2017. "Effects of Polystyrene Microplastics on the Fitness of Earthworms in an Agricultural Soil." IOP Conference Series: Earth and Environmental Science. 61: 012148. https://doi.org/10.1088/1755-1315/61/1/012148.

This paper deals with a very specific issue—the effects of microplastics on earthworms. The authors find that plastics "could inhibit the growth of earthworms and had an obvious lethal effect on earthworms at the high exposure concentrations (0.5%), which were possibly explained by the damage in the self-defense system of earthworms."

Chandler, Nathan. 2020. "How Can We Speed Up Plastic Photo-degradation?" How Stuff Works. https://science.howstuffworks.com/environmental/green-science/speed-up-plastic-photo degradation.htm.

This article provides a good general introduction to the process of photodegradation and its possible use in the production of plastic materials.

Cho, Renee. 2020. "More Plastic Is on the Way: What It Means for Climate Change." State of the Planet. https://blogs.ei.columbia.edu/2020/02/20/plastic-production-climate-change/.

Two important global environmental issues are inextricably bound with each other: plastic production and climate change. This article explores that connection and discusses ways in which greater plastic production is likely to contribute to climate change.

Christiansen, Katie Shepherd. 2018. "Global & Gallatin Microplastics Initiatives." Adventure Scientists. https://www.adventurescientists.org/uploads/7/3/9/8/7398741/2018_microplastics-report_final.pdf.

From 2013 to 2017, Adventure Scientists conducted surveys at 72 locations, where 2,677 water samples were collected to determine the concentration of microplastics present in the samples. This report summarizes the result of that study. It includes an especially useful interactive map of the locations at which samples were collected and the data collected for each location.

"Circular Economy Action Plan." 2020. European Commission. https://ec.europa.eu/environment/circular-economy/pdf/new_circular_economy_action_plan.pdf.

This publication introduces a new program from the European Commission, one based on its belief that the old linear economy approach to products is no longer feasible or desirable. It outlines in detail the elements involved in changing from a linear to a circular economy for many products, plastics prominently among them.

Corkery, Michael. 2019. "A Giant Factory Rises to Make a Product Filling Up the World: Plastic." *The New York Times*. https://www.nytimes.com/2019/08/12/business/energy-environment/plastics-shell-pennsylvania-plant.html.
This story tells of the construction of a giant new fossil fuel cracking plant by the energy company Royal Dutch Shell in Beaver County, Pennsylvania. The plant is envisioned as a way of changing the company's emphasis on fuel production to the manufacture of new plastic products, especially microbeads for personal care, cosmetic, and industrial purposes.

"Defining Life Cycle Assessment." n.d. Global Development Research Center. http://www.gdrc.org/uem/lca/lca-define.html.
Life cycle assessment is a powerful tool for comparing the environmental effects of two or more products, treatments, or other systems. It is useful in comparing the effects of different kinds of plastics with each other or of plastics versus paper, glass, or other materials in the manufacture of products.

Dermansky, Julie. 2020. "The Plastics Giant and the Making of an Environmental Justice Warrior." Desmog. https://www.desmogblog.com/2020/01/07/formosa-sunshine-plastics-sharon-lavigne-environmental-justice.
The environmental justice movement is an effort to call attention to and remedy the trend of industries locating the plants that produce the greatest level of hazardous emissions in neighborhoods occupied primarily by people of color and poor people. This article describes how the problems about which environmental justice is concerned are arising in areas where large new plastic factories are being constructed.

Din, Muhammad Imran, et al. 2020. "Potential Perspectives of Biodegradable Plastics for Food Packaging Application-Review of Properties and Recent Developments." *Food Additives & Contaminants: Part A.* https://doi.org/10.1080/19440049.2020.1718219.

> The authors describe this paper as providing "an insight into recently engineered biodegradable plastics used for packaging applications."

EFSA Panel on Contaminants in the Food Chain (CON-TAM). 2016. "Presence of Microplastics and Nanoplastics in Food, with Particular Focus on Seafood." https://doi.org/10.2903/j.efsa.2016.4501.

> This statement was prepared at the request of the German Federal Institute for Risk Assessment (BfR). It provides some of the most detailed and extensive information on the effects of microplastics and nanoplastics on foods currently available.

"El Dorado of Chemical Recycling: State of Play and Policy Challenges." 2019. Zero Waste Europe. https://zerowasteeurope.eu/wp-content/uploads/edd/2019/08/2019_08_29_zwe_study_chemical_recycling.pdf.

> This study by Zero Waste Europe reviews the role of chemical recycling in a zero waste approach to plastics. It concludes with a series of recommendations as to how this technology can improve the efficiency of a zero waste program for plastics.

"Environmental Justice & the PVC Chemical Industry." n.d. Center for Health, Environment & Justice. http://www.chej.org/pvcfactsheets/Environmental_Justice_and_the_PVC_Chemical_Industry.html.

> This article provides a detailed commentary as to how the production of plastics often can be viewed within the

context of an environmental justice framework. An extensive list of references is available.

"A European Strategy for Plastics in a Circular Economy." 2018. European Commission. https://eur-lex.europa.eu/resource.html ?uri=cellar:2df5d1d2-fac7-11e7-b8f5-01aa75ed71a1.0001.02 /DOC_1&format=PDF.

This document expands on a more general publication, "A European Plan for a Circular Economy," adopted in 2015. It describes in detail the way the recycling and other treatment of plastics can be modified to fit into the broader concept of changing from a make-use-throw away economy into a make-use-remake and reuse economy. It is an excellent general introduction to the concept of circular economy for plastics.

"From a Linear to a Circular Economy." 2017. Government of the Netherlands. https://www.government.nl/topics /circular-economy/from-a-linear-to-a-circular-economy.

One of the most ambitious attempts to use the cradle-to-cradle philosophy of waste management has been adopted by the Dutch government. This website provides an introduction to that plan, along with a document outlining its philosophy and proposed implementation.

"Fueling Plastics: Series Examines Deep Linkages between the Fossil Fuels and Plastics Industries, and the Products They Produce." 2017. Center for International Environmental Law. https://www.ciel.org/reports/fuelingplastics/.

The "Fueling Plastics" project is "an ongoing investigative series, examining the deep linkages between the fossil fuels and plastics industries and the products they produce, and exposing how the U.S. shale gas boom fuels a massive buildout of plastics infrastructure in the United States and beyond." Among the reports produced as part of that program are "Fossils, Plastics, and Petrochemical Feedstocks,"

"How Fracked Gas, Cheap Oil, and Unburnable Coal Are Driving the Plastics Boom," "Plastic Industry Awareness of the Ocean Plastics Problem," and "Untested Assumptions and Unanswered Questions in the Plastics Boom."

Fuhr, Lili, and Matthew Franklin. 2019. "Plastic Atlas: Facts and Figures about the World of Synthetic Polymers." Heinrich Böll Foundation and Break Free from Plastic. 2nd ed. https://www.boell.de/sites/default/files/2020-01/Plastic%20Atlas%202019%202nd%20Edition.pdf.

This excellent publication consists of 19 heavily illustrated chapters that are aimed at providing "in-depth knowledge of the causes, interests, responsibilities and effects of the plastics crisis."

Geyer, Roland, and Jenna Jambeck. 2017. "Production, Use, and Fate of All Plastics Ever Made." *Science Advances*. 3(7): e1700782. https://doi.org/10.1126/sciadv.1700782.

The authors attempt to estimate the total amount of plastic ever made and the fate of that plastic. They estimate that 8,300 million metric tons of plastic were made prior to 2017. of which 9 percent was recycle, 12 percent was incinerated, and 79 percent was left in landfills or the natural environment.

Goldsberry, Clare. 2019. "France Charts Course for Zero-waste Society." Plastics Today. https://www.plasticstoday.com/packaging/france-charts-course-zero-waste-society/190772706162110.

This article discusses France's decision to ban all plastic packaging by the year 2040.

"A Guide to Plastic in the Ocean." 2019. National Ocean Service. https://oceanservice.noaa.gov/hazards/marinedebris/plastics-in-the-ocean.html.

This relatively short introduction to plastics pollution of the oceans is especially useful because of the many links it provides to other articles and Internet sources dealing with specific topics on the subject.

Hamilton, Lisa Anne, and Steven Feit. 2019. "Plastic & Climate: The Hidden Costs of a Plastic Planet." Center for International Environmental Law, et al. https://www.ciel.org/wp-content /uploads/2019/05/Plastic-and-Climate-FINAL-2019.pdf.
This report discusses in detail the many ways in which plastic contribute to global climate change, from extraction and transport of raw materials to refining and manufacture to use and disposal. The authors claim that greenhouse gas emissions from this whole process are likely to make it impossible for humans to keep the global average annual temperature from increasing by 1.5 C.

Holden, Emily. 2019. "Will a Push for Plastics Turn Appalachia into next 'Cancer Alley'?" The Guardian. https://www .theguardian.com/environment/2019/oct/11/plastics-appalachia -next-cancer-alley-fracking-public-health-ethane.
The success of fracking operations in the United States has produced an abundance of natural gas for which there is now no corresponding demand. One way of using the product is by cracking it to make the monomers of which plastics are made. That process holds the potential, however, for resulting in a substantial release of harmful chemicals to the surrounding countryside.

Howe, Angela, et al. 2019. "Federal Actions to Address Marine Plastic Pollution: Reducing or Preventing Marine Plastic Pollution through Source Controls and Life-cycle Management." UCLA School of Law Emmett Institute. Surfrider Foundation. https:// law.ucla.edu/centers/environmental-law/emmett-institute-on -climate-change-and-the-environment/publications/federal -actions-to-address-marine-plastic-pollution/.
This report acknowledges the beginnings of federal efforts to deal with the nation's and world's plastic pollution

problems. It assesses current steps in that direction and recommends other federal actions that can be effective in reducing this problem.

"IMO Committed to Plastic Waste Regulation." 2016. International Marine Organization. https://www.maritime-executive.com /article/imo-committed-to-plastic-waste-regulation.
A variety of international treaties and agreements contain provisions for the control of plastic debris dumped into the oceans. This article reviews provisions of some of these agreements.

Joyce, Christopher. 2019. "Where Will Your Plastic Trash Go Now That China Doesn't Want It?" NPR. https://www.npr.org /sections/goatsandsoda/2019/03/13/702501726/where-will -your-plastic-trash-go-now-that-china-doesnt-want-it.
This article discusses the history of China's involvement with the world's solid waste disposal and the problems created by its decision not to accept those wastes any longer.

"Leo Hendrick Baekeland and the Invention of Bakelite." 1993. American Chemical Society. https://www.acs.org/content /acs/en/education/whatischemistry/landmarks/bakelite.html #biography-of-leo-baekeland.
This booklet/web page provides a nice general introduction to the "Age of Plastics," as well as history of the invention of Bakelite plastic and a biography of its inventor, Leo Baekeland.

Lerner, Sharon. 2019. "Leaked Audio Reveals How Coca-Cola Undermines Plastic Recycling Efforts." The Intercept. https://theintercept.com/2019/10/18/coca-cola-recycling -plastics-pollution/.
This article is based on a supposed private telephone conversation among recycling leaders in Atlanta that includes a commentary about efforts by the Coca-Cola Company

to reduce its dependence on plastic packaging materials. Also see "Sustainable Packaging 2019."

"Life Cycle Assessment for a Plastic and a Glass Product." 2013. Life Cycle Assessment. https://lifecycleofplastic.wordpress.com /2013/10/08/life-cycle-assessment-for-a-plastic-and-a-glass -product/.
    Life cycle assessment is an analytical system that compares two or more systems of some kind with regard to their environmental effects at all stages of extraction, transportation, production, consumer use, and disposal. This case study illustrates the use of this methodology in comparing relative effects of using plastic and glass products of the same general type.

Liggat, John. n.d. "ICI's Biopol. Cautionary Tales." University of Strathclyde. https://www.scotchem.ac.uk/wp-content/uploads /2019/02/Biopol-IBioIC-compressed.pdf.
    The author provides an extensive and detailed account of the first commercially available biodegradable plastic, Biopol. The article includes a history of research on the subject, relevant physical and chemical properties, possible applications, and potential environmental and other problems associated with the material.

Machado, Anderson Abel de Souza, et al. 2017. "Microplastics as an Emerging Threat to Terrestrial Ecosystems." *Global Change Biology*. https://doi.org/10.1111/gcb.14020.
    One of the less well-studied topics in the field of microplastics is how those materials affect terrestrial organisms and ecosystems. This article reviews current knowledge in the field and outlines some recommended areas of further research.

"Making Plastics: Extracting Raw Materials." n.d. Paprec Group. https://www.paprec.com/en/understanding-recycling /recycling-plastic/making-plastic-extracting-raw-material.

The production of plastics has environmental effects at every stage of the operation. This article discusses some of the effects that occur in the very first stage—extraction of raw materials from the earth.

"Microplastics in the Environment." 2019. Society of Environmental Toxicology and Chemistry. https://setac.onlinelibrary .wiley.com/doi/toc/10.1002/(ISSN)1551-3793.microplastics.

This special issue was produced to support a webinar on the topic. It includes chapters on current understanding of microplastics in the environment, impacts of ocean circulation on microplastic distribution, microplastics as contaminants in seafood products, microplastics in the aquatic environment, and microplastics in freshwater environments.

"New Plastics Economy." 2017. Ellen MacArthur Foundation. https://www.newplasticseconomy.org/.

This website provides an introduction to a new program for re-imagining the way plastics are produced, used, and recycled today. It is an ambitious and very optimistic approach to the problem of plastic wastes in the world.

Oakes, Kelly. 2019. "Why Biodegradables Won't Solve the Plastic Crisis." BBC Future. https://www.bbc.com/future/article /20191030-why-biodegradables-wont-solve-the-plastic-crisis.

Many people are hopeful that biodegradable plastics may be an answer to the problem of plastic pollution today. But this article points out that that may be true for only certain types of biodegradables. The author reviews the topic in general and explains the prospective value of various types of biodegradables.

"Our New Initiative: Together, We're Committed to Getting Every Bottle Back." 2019. American Beverage Association. https://www.innovationnaturally.org/plastic/.

This joint program of the Coca-Cola Company, Keurig Dr. Pepper, and Pepsico in association with the World Wildlife Fund, Closed Loop Partners, and The Recycling Partnership is an effort to reduce the amount of plastic wastes resulting from the use of their products. It involves the use of especially recyclable plastic materials designed to be returned to and reused by the beverage companies.

Peters, Adele. 2019. "The World's Big Plastic Makers Want More Recycling So They Can Keep Pumping out Plastic." Fast Company. https://www.fastcompany.com/90295292/the -worlds-big-plastic-makers-want-more-recycling-so-they-can -keep-pumping-out-plastic.

A group of plastic-making and fossil fuel companies have recently initiated a campaign to eliminate plastic wastes worldwide (see Together, We Can End Plastic Waste Forever 2020). This article raises questions about the sincerity and possible effectiveness of this campaign in the reduction of plastic wastes.

"Planet or Plastics." 2020. National Geographic. https://www .nationalgeographic.com/environment/planetorplastic/.

This page provides an introduction to the organization's "multiyear effort to raise awareness about the global plastic trash crisis." It contains links to dozens of articles on specific aspects of the plastics pollution problem, including the role of toothbrushes in the problem, bans by states and national governments on plastic use, the special problems of plastic wraps, consumption of microplastics by corals, plastics in the Gaza strip, and the occurrence of microplastics in the Arctic.

"Plastic in Cosmetics." n.d. United Nations Environment Programme. http://wedocs.unep.org/bitstream/handle/20.500.11822 /21754/PlasticinCosmetics2015Factsheet.pdf.

This fact sheet provides a good overview of the role of primary microplastics in cosmetics, with a list of the most commonly used materials and the purposes for which they are employed in the field.

"Plastic Industry Daily News and More . . ." 2019. plastics .com.
This website provides a comprehensive review of a wide range of plastic-related topics, many of which are of special interest to the industry itself.

"Plastic Packaging Tax." 2020. Department for Environment Food & Rural Affairs. https://consult.defra.gov.uk/environmental -quality/plastic-packaging-tax/.
This governmental document explains the imposition of a new tax on plastic packaging products in the United Kingdom, effective April 2022. The tax will amount to £200 per metric ton of packaging made from less than 30 percent recycled plastic.

"Plastics Hall of Fame." 2020. https://www.plasticshof.org/.
This association was formed to honor individuals who have made special contributions to the development of the modern plastic industry. It contains more than 400 biographical sketches of important women and men in the field.

"Polyethylene Terephthalate (PET): A Comprehensive Review." 2019. Omnexus. https://omnexus.specialchem.com /selection-guide/polyethylene-terephthalate-pet-plastic.
This website provides an excellent and easily understandable introduction to one of the most popular of all plastics, polyethylene terephthalate (PET).

Prata, Joana Correia, et al. 2020. "Environmental Exposure to Microplastics: An Overview on Possible Human Health

Effects." *Science of the Total Environment.* 702: 134455. https://doi.org/10.1016/j.scitotenv.2019.134455.

> The authors survey the topic of potential human health effects from exposure to microplastics and report on possible sources, methods of transmission, and research evidence of possible effects.

Ragaert, Kim. 2016. "Trends in Mechanical Recycling of Thermoplastics." Conference Paper: Kunststoff Kolloquium Leoben. Research Gate. https://www.researchgate.net/publication/302583613_Trends_in_Mechanical_Recycling _of_Thermoplastics.

> This paper provides a somewhat technical review of one of the major types of plastic recycling, mechanical recycling, as used with thermoplastic materials.

Rankin, Jennifer. 2020. "EU Plans 'Right to Repair' Rules for Phones and Tablets." The Guardian. https://www.theguardian .com/world/2020/mar/11/eu-brings-in-right-to-repair-rules-for -phones-and-tablets.

> As part of its Circular Economy Action Plan, the European Commission is encouraging manufacturers to design and build products intended for reuse rather than disposal. The main elements of this plan are to have companies make available spare parts and repair services for their products, along with free or inexpensive upgrading of their products.

Reusch, William. 1999. "Polymers." Virtual Textbook of Organic Chemistry. https://www2.chemistry.msu.edu/faculty /reusch/VirtTxtJml/polymers.htm.

> This web page provides a good, general, and somewhat advanced introduction to all aspects of polymeric chemistry and their applications to plastics.

Ryan, V. 2014. "Biopol — Biodegradable Plastic." Technologystudent.com. http://www.technologystudent.com /prddes1/biopola.html.

This article describes the first biodegradable plastic invented and made commercially available by the British chemical company, ICI, in 1990.

"Science-Based Solutions to Plastic Pollution." 2020. *One Earth*. 2(1): 5–7. https://doi.org/10.1016/j.oneear.2020.01 .004.

The "Voices" column of this journal asks a variety of researchers how science can contribute to the solution of the world's plastic pollution problem. Some suggestions offered include a circular plastics economy, biological systems, wise use of aquaculture, and better informed action by ordinary people.

Seldman, Neil. 2019. "Federal Actions on Plastic Recycling, Zero Waste, Organics, and Garbage Incineration: A Review." Institute for Local Self-Reliance. https://ilsr.org /review-of-federal-actions-on-plastic/.

The author provides a review of proposed federal legislation on plastic wastes and associated issues, along with a critique of each proposal.

Shaw, Emma J., and Andrew Turner. 2019. "Recycled Electronic Plastic and Marine Litter." *Science of the Total Environment*. 694: 133644. https://doi.org/10.1016/j.scitotenv.2019.133644.

The authors pursue an intriguing question as to how the color of discarded plastic materials can affect the character of plastic wastes that end up in the oceans. They find that black wastes from electrical and electronic equipment contribute measurable amounts of toxic chemicals, such as antimony, bromine, cadmium, chromium, and lead, to marine organisms.

Siegler, Kirk. 2019. "On the Oregon Coast, Turning Pollution into Art with a Purpose." NPR. https://www.npr.org/2019/12 /04/784416386/on-the-oregon-coast-turning-pollution-into -art-with-a-purpose.

The Washed Ashore project described here was created in 2010 for the purpose of collecting waste plastics on the Oregon coast and then using those materials for the construction of sculptures and other art objects.

Sobczyk, Nick. 2020. "Environmental Justice Gets Its Day in the Sun." E&E News. https://www.eenews.net/stories /1062221805.

The author explains how the construction of a large new plastics plant in Louisiana poses familiar environmental justice issues for residents of the area.

"Stephanie L. Kwolak." n.d. Science History Institute. https:// www.sciencehistory.org/historical-profile/stephanie-l-kwolek.

In 1965, Kwolek invented Kevlar, a material so strong and resistant that it was first used for the production of bulletproof vests. It has since found more than 200 other applications.

"Sustainable Packaging." 2020. The Coca-Cola Company. https://www.coca-colacompany.com/sustainable-business /packaging-sustainability.

Certain industries have contributed to problems of plastic waste accumulation to a significant degree. Bottling companies that produced bottled water, sodas, fruit drinks, milk products, and other liquid products are among these industries. As the problem of plastic waste increases and consumers begin to raise issues about the use of plastic packaging, many companies have begun to distance themselves from the use of plastics and turn to more traditional packaging materials, such as glass bottles. This web page describes the Coca-Cola Company's position on this issue and its plans for reducing its dependence on plastic packaging materials. Also, see the view of some environmental groups about efforts such as these at Lerner 2019.

Taylor, Matthew. 2017. "$180bn Investment in Plastic Fac-
tories Feeds Global Packaging Binge." The Guardian. https://
www.theguardian.com/environment/2017/dec/26/180bn
-investment-in-plastic-factories-feeds-global-packaging-binge.
   Fossil fuel companies are now investing huge amounts of
   money in the construction of new plants for the production of plastics, exacerbating plastic use and plastic waste
   problems that are now endemic across the planet.

Thomlinson, Isabel. 2019. "When Biodegradable Plastic Is Not
Biodegradable." The Conversation. https://theconversation.com
/when-biodegradable-plastic-is-not-biodegradable-116368.
   The author explains the differences between various types
   of "biodegradable" plastics and comments on a study
   showing that some forms do not degrade very well at all
   in natural environments. For the study itself, see Napper
   and Thompson under "Articles."

"Together, We Can End Plastic Waste Forever." 2020. Alliance
to End Plastic Waste. https://endplasticwaste.org/.
   Among the founding members of this group are chemical and energy companies Chevron Phillips Chemical
   Company, Dow Chemical, ExxonMobil, Formosa Plastics Corporation, Mitsubishi Chemical Holdings, Mitsui
   Chemicals, NOVA Chemicals, Procter & Gamble, Reliance Industries, Shell, Sumitomo Chemical, Total, and
   Veolia. The organization expects to spend over $1.5 billion
   over the next five years to solve the world's plastic waste
   problems. Some questions have been raised, however,
   regarding the effectiveness of some of the world's largest
   plastic-making companies to make significant progress in
   this goal. See, for example, Peters 2019.

"12 Inspiring Works of Art on Plastic Pollution." 2017. Plastic Pollution Coalition. https://www.plasticpollutioncoalition

.org/blog/2017/5/2/10-inspiring-works-of-art-about-plastic
-pollution.

Many artists have found plastic wastes to be an inspiring way of producing sculptures, paintings, and other works of art. To some extent, these efforts are aimed at helping in reducing the amount of plastic waste in the environment. But they are also intended to increase the public's understanding of the world's plastic waste crisis.

Valentine, Eric M. 2018. *Plastic Pollution in Rivers and Oceans.* Marlow, Bucks, UK: Foundation for Water Research. http:// www.fwr.org/environw/frr0030.pdf.

This short book attempts to offer the most up-to-date information available on the subject of plastic pollution in the oceans. Topics include types of plastics and their uses, sources of plastic waste, persistence in the oceans, spread of plastic wastes, impact of plastics, and appropriate changes needed by humans.

van Odijk, Sanderine, and Anne Poggenpohl. 2019. "Going Plastic Free: What Does a Zero Waste Future Look Like?" Greenpeace. https://www.greenpeace.org/usa/zero-waste-future/.

This article outlines the principal concepts involved in a zero waste program of plastics. The podcast is the second in a series called "Going Plastic Free," the first of which can be found at https://www.greenpeace.org/usa /myths-about-ending-plastic-pollution/.

"Welcome to the Plastics Historical Society." 2015. Plastics Historical Society. http://plastiquarian.com/.

This organization is a treasure chest of information about many aspects of plastics, including their history in society. Unfortunately, most items are available to paid members only. However, some good sections are available to the general public, such as "Plastics in Society" and "People & Polymers."

## Introduction

The world is filled with natural polymers, of which plastics are one form. Any history of plastics could easily go back to the days of ancient China and Egypt, where humans were already making use of natural plastics, most commonly some form of the cellulose that occurs in all plants. The modern age of plastics dates, however, only to the nineteenth century, when many of the most basic facts about polymeric substances were developing. Indeed, many authorities date the start of the modern age of plastics only to 1907, when Belgian chemist Leo Baekeland invented a plastic he called *Bakelite*. This chronology, then, focuses on the most important discoveries and other events in the field of plastics that have occurred over the past 300 years or so. The exact dates for many events are often disputed among experts. The "correct" dates, however, are often within a few years of each other.

**1496**   Christopher Columbus returns from a journey to the New World (South America), bringing with him the first samples of natural rubber to Europe. The product had been known and used for centuries by Aztecs, from whom Columbus obtained his samples. The material was used, among other purposes, to make rubber balls for their traditional game of ollama. Aztecs referred to the product as *caoutchouc*, for "weeping wood."

---

The way in which plastics are used in the manufacture of tools, toys, sporting equipment, and other everyday objects is probably more than can be accurately estimated. (James Copeland/Dreamstime.com)

**1656**   English natural historian John Tradescant is thought to be the first European to have discovered gutta percha, the tough plastic material obtained from rubber trees.

**1820s–1843**   The uses of natural rubber are limited because of some limiting natural properties. Researchers, primarily in Great Britain and the United States, search for ways to improve on these properties to make possible a wider range of applications for natural rubber. Thomas Hancock in England and Charles Goodyear in the United States almost simultaneously develop a method for hardening ("vulcanizing") rubber. The new product is named *ebonite* or *vulcanite*.

**1831**   French chemist Jean-François Bonastre extracts a resin from the balsam tree, *Liquidambar storax*, in which he finds small amounts of a new compound he calls *styrene*. The compound is later to be used as a monomer in the production of polystyrene plastics.

**1838**   French physicist Henri Victor Regnault discovers vinyl chloride, whose molecule structure is identical to that of ethylene, with one chlorine substituted for one hydrogen in the molecule. Regnault does not follow up on his discovery, but does note that the compound changes (polymerizes) to a white solid when exposed to sunlight.

**1843**   English physician William Montgomerie, then serving as assistant surgeon to the president of Singapore, discovers that gutta percha is a useful product for the manufacture of some types of medical instruments. This event is the first occasion on which the substance finds practical uses in society.

**1843**   Austrian chemist Joseph Redtenbacher first synthesizes acrylic acid, the basic monomer on which the class of plastics known as the polyacrylates were later invented.

**1856**   An early type of plastic known as *bois durcit* is invented by the French poet Francois Charles Lepage. The product was made of blood and powdered wood and then treated until it had the physical properties of a plastic, wood-like material.

**1856 (Some sources cite 1862)**    English chemist Alexander Parkes invents a plastic-like product made by dissolving nitrocellulose in alcohol or wood naphtha. When mixed with vegetable oil or camphor, the product polymerizes to form a hard, tough, flexible horn-like material with many applications. He calls the product *parkesine*. Parkes's partner, Daniel Spill, adapts his recipe to produce a very similar (if not identical) product called *xylonite*. Some historians credit Parkes with inventing the first celluloid plastic, although that encomium is also given to John Wesley Hyatt.

**1865**    German chemist Paul Schützenberger invents cellulose acetate, a compound formed in the reaction between natural cellulose and acetic anhydride, a form of vinegar. The compound finds little practical use until a practical method of production is developed in 1894 by Charles Frederick Cross and Edward John Bevan.

**1869**    American inventor John Wesley Hyatt invents celluloid, often called the first true plastic introduced to the world. The product is made from a mixture of cellulose nitrate, camphor, and alcohol. It is easy to work with and converted to a variety of shapes and forms. The product soon becomes the raw material from which dozens of consumer products, from hair combs to billiard balls, are made.

**1872**    John Wesley Hyatt and his brother, Isaiah, receive a patent for the world's first injection molding machine. A critical tool in today's plastic industry, the machine worked by injecting a liquid plastic into a mold of some shape or another and then allowed to dry, producing a product of the desired size and shape.

**1874**    British engineer Albert Freyer receives a patent for the first modern incinerator, a furnace in which solid wastes could be burned.

**1880**    Swiss chemist Georg W. A. Kahlbaum synthesizes the first polymer of acrylic acid, polymethylacrylate.

**1885**    American inventor George Eastman invents the first transparent photographic film, made out of cellulose nitrate that could be wound on a spindle. The product greatly improves the speed and fidelity of motion picture products.

**1892**    English chemists Charles Frederick Cross and Edward John Bevan receive a patent for the production of viscose plastic. The material is made of cellulose fiber obtained from wood and other agricultural fibers. Viscose is later used as the basis for the production of rayon, a product known as *viscose rayon*.

**1894**    Cross and Bevan receive a patent for a method for making cellulose acetate plastic. The substance then becomes widely popular for the manufacture of a variety of goods, such as photographic film, magnetic tape, and fibers. Cellulose acetate is sometimes regarded as a type of rayon.

**1897**    Bavarian chemist Adolf Spitteler accidentally discovers that a hard, attractive plastic can be made by mixing formaldehyde with ordinary milk. In the reaction, the protein in milk called *casein* undergoes a polymerization reaction that results in the formation of plastic casein.

**1901**    German chemist Otto Karl Julius Röhm writes his doctoral dissertation on the polymerization of acrylic acid, vastly increasing the basic knowledge about the process. The reaction was not developed sufficiently for commercial use for almost three decades more (see **1928**).

**1907**    Belgian chemist Leo Baekeland invents a phenol-formaldehyde polymer type of plastic that he calls *Bakelite*. The invention is sometimes said to be the beginning of the Age of Plastics.

**1912**    German chemist Fritz Klatte discovers the plastic polyvinyl acetate (today, also known as PVA). The compound is produced in the reaction between two simple organic compounds, acetylene ($C_2H_2$) and acetic acid ($C_2H_4O_2$).

**1918**    Czech chemist Hanns John obtains a patent for the production of a class of plastic polymers known as *urea-formaldehyde resins.* They are made by way of condensation reactions between urea ($CH_4N_2O$) and formaldehyde ($CH_2O$).

**1928**  German chemists Otto Röhm and Otto Haas begin production of a commercial version of poly(methyl acrylate) under the names of Acryloid and Plexigum. The product is recommended as a binder between two sheets of glass in the form of "safety glass" much in demand in the automotive industry.

**1929**  Chemists at the I.G. Farbenindustrie German chemical company develop methods for the production of polystyrene from the monomer styrene (see **1831**). Almost a century was needed to find a method for making the product, partly because it polymerizes so quickly under ambient conditions.

**1930**  American chemist Wallace Hume Carothers discovers the principle behind the production of polyesters. He hypothesizes that reactions between diols (compounds with two -OH groups) and diacids (compounds with two COOH groups) would occur, resulting in the loss of water in condensation reactions and the formation of long chains of ester linkages. He successfully confirmed this hypothesis, calling the new polymers *polyesters*. He does not pursue the commercial development of the product, however, because of his competing research on nylon. (See next entry.)

**1930**  Carothers (see the preceding entry) conducts research on the development of polyesters. In this case, he explores the reaction between diamines (compounds with two $NH_2$ groups) and diacids. The polymers formed consist of many repeating amide (RNHCOR') linkages and are called *polyamides*. The most famous of these polyamids are a group of compounds known as *nylon*, the best known of which are *nylon-6* and *nylong-66*.

**1933**  American inventor, Waldo Semon, invents polyvinyl chloride. He is originally hired by the B. F. Goodrich company to develop an inexpensive substitute for natural rubber. But he leaves that field to investigate the polymerization of vinyl chloride, finding that, under the proper conditions, a flexible, waterproof type of sheeting material is formed. The product quickly becomes widely popular for the production of waterproof garments and for water-resistant sheeting needed in commercial and industrial operations.

**1933**   Working in the laboratories at the Imperial Chemical Company in London, British chemists Reginald Gibson and Eric William Fawcett accidentally discover low-density polyethylene. Their original research involves a totally different polymerization reaction between ethylene and benzaldehyde. But they notice the formation of a heavy, waxy, white solid every time they attempt to carry out the polymerization. They later identify the product as polyethylene, specifically low-density polyethylene. It takes more than two years for them to develop a commercially viable polyethylene for producing plastic, which soon becomes very popular for a variety of applications.

**1933**   British chemists John Crawford and Rowland Hill develop a new type of plastic called polymethyl methacrylate. The substance has many of the qualities of glass, along with a greater resistance to breakage. It eventually finds extensive use in the form of the product called Plexiglas®.

**1938**   American chemist Roy J. Plunkett accidentally discovers that polymerization of tetrafluoroethylene forms polytetrafluorene. The compound eventually becomes a huge commercial success, most commonly in the form of the plastic covering called Teflon®.

**1949**   The first hydraulic fracturing ("fracking") operations are carried out in the United States. The success of the technology is that it later makes possible the recovery of very large amounts of ethane and other fossil-fuel gases, which, in turn, becomes a robust source of monomers for plastic production.

**1965**   Swedish engineer Sten Sustaf Thulin invents a one-piece polyethylene shopping bag, patented under the name of

Celloplast. The plastic bag quickly begins to replace cloth and plastic in many parts of Europe.

**1965**   American chemist Stephanie L. Kwolek invents a new type of plastic, Kevlar, that is unusually strong and stiff. It is eventually used in applications ranging from bulletproof vests to boats to ropes and cables.

**1967**   The movie *The Graduate* includes one of the most famous quotations in film history. When giving advice to young college graduate Benjamin Braddock, his older business-man friend, Mr. McGuire, in advising a career choice, says, "I want to say just one word to you. One word." That word, he goes on, is "plastics."

**1972**   The world's major maritime nations adopt the London Convention on the Prevention of Marine Pollution by Dumping of Wastes and other Matter. The agreement includes mention of, but little emphasis on, the role of plastic materials in the world's oceans' dumping issues.

**1972**   Researchers at the Woods Hole Oceanographic Institute report the discovery of "spherules" made of polystyrene in the coastal waters of southern New England. They identify two types of particles averaging 0.5 millimeters in size: one a clear, crystalline form and the other a white, opaque form consisting of synthetic rubber.

**1972**   In what might be the first known use of the material, the Unilever Company obtains a patent for the use of micro-beads as exfoliants in personal care products.

**1976**   Economists Walter Stahel and Genevieve Reday offer the first modern theory of a circular economy in their paper, "Potential for Substituting Manpower for Energy."

**1979**   Plastic shopping bags are introduced to the United States.

**1980**   The U.S. Congress adopts The Act to Prevent Pollution from Ships in order to bring the United States into agreement with the International Convention for the Prevention of

Pollution from Ships (MARPOL). The act is later amended by the Marine Plastic Pollution Research and Control Act of 1987.

**1982** Two of the United States' largest supermarket chains, Kroger and Safeway, switch from paper to plastic bags at checkout stands.

**1984** The Workshop on the Fate and Impacts of Marine Debris is held in Honolulu. The purpose of the meeting is to "bring together fishery biologists, fishery scientists, oceanographers, some population modeling types, and some folks in the fishing industry in an attempt to make an honest assessment of whether marine debris, particularly derelict fishing gear, was a problem worth people's attention" (McIntosh, Naomi, et al. 2000. "Proceedings of the International Marine Debris Conference On Derelict Fishing Gear and the Ocean Environment." NOAA. National Marine Sanctuaries. https://nmshawaiihumpbackwhale.blob.core.windows.net /hawaiihumpbackwhale-prod/media/archive/documents/pdfs _conferences/proceedings.pdf). Later conferences in the series are held in 1989, 1994, 2000, 2011, and 2018.

**1987** The U.S. Congress adopts the Marine Plastic Pollution Research and Control Act, amending the Act to Prevent Pollution from Ships of 1980 to include (among other actions) specific prohibitions against the disposal of plastic debris in U.S. coastal waters.

**1988** The Shore Protection Act is adopted to prevent and reduce the disposal or loss of harmful materials from waste-transporting ships along the nation's coastal waterways.

**1988** Annex V of the International Convention for the Prevention of Pollution from Ships, 1973 as modified by the Protocol of 1978 (MARPOL 73/78) comes into effect. A major component of this part of the MARPOL agreement is a complete ban on the dumping of plastics in the oceans by ships at sea.

**1988** The Marine Protection, Research, and Sanctuaries Act has three major components: prohibition of transporting wastes from the United States for the purpose of ocean dumping; prohibition of transporting wastes from anywhere in the world on U.S.-flagged ships for the purpose of ocean dumping; and prohibition of the transport of wastes for dumping in territorial seas. The act is often referred to as the Ocean Dumping Act.

**1989**   The Second International Marine Debris Conference is held in Hawaii.

**1990**   Researchers at Great Britain's ICI chemical manufacturing company invent a biodegradable type of plastic, polyhydroxybutyrate, sold commercially as Biopol.

**1994**   The Third International Marine Debris Conference is held in Miami.

**1997**   Marine researcher Charles Moore discovers the Great Pacific Garage Patch in the central north Pacific ocean.

**2000**   The Beaches Environmental Assessment and Coastal Health (BEACH) Act is adopted as an amendment of the Clean Water Act in 2000. Its purpose is to prevent the dumping of hazardous wastes in recreational waters adjacent to the U.S. coastline.

**2000**   The Fourth International Marine Debris Conference is held in Miami.

**2002**   Bangladesh becomes the first country in the world to ban plastic bags. The action is taken because disposed plastic bags blocked sewers nationwide during the traditional flooding season.

**2002**   American engineer Richard Anthony develops the concept of a zero waste economy, with special significance for its application to the problem of plastic wastes.

**2002**   In their book *Cradle to Cradle: Remaking the Way We Make Things*, architect William McDonough and chemist Michael Braungart outline the basic principles of the modern theory of circular economy, defined as a system for designing out waste and pollution, keeping products and materials in use, and regenerating natural systems.

**2004**   British marine biologist Richard C. Thompson is credited with having first used the term *microplastics* in the academic literature. The term appears in his article, "Lost at Sea: Where Is All the Plastic?" *Science*. 304(5672): 838.

**2006**   The U.S. Congress passes the Marine Debris Research, Prevention, and Reduction Act. The act created a Marine Debris Program within NOAA and provided funds for programs on the reduction and prevention of adverse effects resulting from ocean pollutants.

**2007**   San Francisco becomes the first city in the United States to ban all use of plastic bags in the city.

**2009**   The Boeing aircraft company brings into service its new 787, sometimes called "Boeing's Plastic Dream." The aircraft skin consists of 100 percent plastic materials, while about half of all the materials used overall are plastic.

**2011**   Experts estimate that consumers worldwide are consuming plastic bags at the rate of one million bags per minute.

**2011**   The Fifth International Marine Debris Conference is held in Honolulu.

**2013**   Eighteen-year-old Dutch high school boy Boyan Slat conceives the idea of building a system for removing plastic wastes from the Great Pacific Garbage Patch. He forms a company, Ocean Cleanup, to design the necessary equipment and test and put it into operation in the region.

**2014**   The first United Nations Environment Assembly (UNEA) of the United Nations Environment Programme meets in Nairobi, Kenya, and adopts 17 resolutions on various aspects of the planet's environmental health. One such document is known as Annex I of Resolution1/6, which states the organization's position on plastic debris in the world's oceans. That document contains a very detailed and comprehensive summary of essentially all that is known about the occurrence, concentration, and effects of plastics and microplastics in the oceans and some freshwater environments.

**2015**  The Microbead-Free Waters Act of 2015 prohibits the addition of microplastic beads to certain so-called "rinse-off" personal care products, such as toothpaste and some cosmetics. The purpose of the act is to prevent the release of microplastics into the nation's freshwater resources and the oceans.

**2015**  A review chapter on nanoplastics contains what is presumably the first formal use of the term in the scientific literature (Albert A. Koelmans, Ellen Besseling, and Won J. Shim. "Nanoplastics in the Aquatic Environment. Critical Review." In Melanie Bergmann, Lars Gutow, and Michael Klages, eds. *Marine Anthropogenic Litter.* 325–340. Cham, Switzerland: Springer).

**2016**  The World Economic Forum and the Ellen MacArthur Foundation launch a new program to deal with plastic pollution in the oceans, The New Plastics Economy—Rethinking the Future of Plastics.

**2017**  The United Nations Environment Program launches #CleanSeas, a social media program designed to educate the general public and the private sector about the problem of plastic wastes in the world's oceans.

**2018**  The Ocean Cleanup company conducts its first tests of System 001 for the cleanup of sections of the Great Pacific Garbage Patch. The test is unsuccessful. A modified and improved version of the system is tested successfully in late 2019.

**2018**  The Sixth International Marine Debris Conference is held in San Diego.

**2018**  Seattle becomes the first city in the United States to ban plastic straws and utensils.

**2018**  Chile's Constitutional Court approves a recent law banning the retail use of plastic bags throughout the country.

**2018**  The Walt Disney Company announces a ban on the use of single-use plastic straws and stirrers at nearly all of its theme parks and resorts.

**2018**   The Danish brewing company, Carlsberg, announces discontinuation of the use of plastic multipack rings, the first major beer company to adopt that policy.

**2018**   President Donald Trump signs the Save Our Seas act to reduce the amount of plastic pollution in the oceans.

**2018**   The Supreme Court of Texas strikes down an ordinance adopted by the city of Laredo reducing the use of one-time paper and plastic bags in the city. The court rules that the city ordinance preempts state law on the topic of waste management.

**2018**   China bans the import of plastic wastes from other countries, including the United States.

**2019**   The city of San Diego bans the use of Styrofoam food and drink containers in stores, restaurants, and other food outlets.

**2019**   Peru bans the use of all single-use plastic objects in its 76 cultural and natural protected areas.

**2019**   Ocean Cleanup successfully tests System 002/B in a cleanup of a portion of the Great Pacific Garbage Patch. (Also see **2018**.)

**2019**   In the case of *San Antonio Bay Estuarine Waterkeeper and S. Diane Wilson vs. Formosa Plastics Corp., Texas, and Formosa Plastics Corp*, Judge Kenneth M. Hoyt of the District Court for the Southern District of Texas, Victoria Division, finds in favor of the plaintiffs and approves a settlement requiring the company to provide compensation and changes in policy to prevent pollution of the bay by its products in the future.

**2020**   The British government announces the implementation of a new Plastics Packaging Tax to discourage the use of plastics for packaging purposes. The tax will amount to £200 per metric ton of packaging made from less than 30 percent recycled plastic. It is scheduled to go into effect in April 2022.

**2020**   China's National Development and Reform Commission and Ministry of Ecology and Environment issue a new policy banning plastic bags in all of China's major cities by the end of 2020 and in all cities and towns in 2022. Markets selling fresh produce are exempt from the ban until 2025.

**2020**   Malaysia returns 150 containers of plastic wastes to 13 countries, confirming that it would no longer accept such materials for recycling. The wastes were sent back to France (43 containers), United Kingdom (42), United States (17), Canada (11), Spain (10), and the rest to Bangladesh, China, Hong Kong, Japan, Lithuania, Portugal, Singapore, and Sri Lanka.

**2020**   U.S. Senators Tom Udall (D-NM) and Jeff Merkley (D-OR) and U.S. Representatives Alan Lowenthal (D-CA) and Katherine Clark (D-MA) unveil the Break Free From Plastic Pollution Act of 2020. The legislation is designed to phase out unnecessary single-use plastic products, hold corporations accountable for wasteful products, reduce wasteful packaging, and reform the nation's waste and recycling collection system.

# Glossary

As with any other field of science, the study of plastics and microplastics makes use of a substantial number of technical terms for an understanding of the topic. Most of these terms have already been mentioned or defined in the main text of this book. But many of the terms have not been used here, but are defined to assist the reader with her or his further research on and the understanding of the subject. Some terms have applications in a variety of fields. The understanding here is that the definition applies specifically to the field of plastics and microplastics.

**abrasion resistance** The ability to withstand mechanical actions on the surface of a material without deforming the shape of that surface.

**accelerator** A substance that is added to the reaction involving a plastic resin and a catalyst, designed to increase the rate at which the reaction occurs. It is used most commonly in the production of rubber, also: **promoter**.

**acrylate resin** A plastic material formed by the polymerization of acrylic acid or one of its derivatives.

**acrylic acid** An organic compound with the chemical formula CH=CHCOOH.

**additive (plastic)** A substance that is added to a polymer to give it some specialized property, such as color or strength.

**advanced recycling** *See* **chemical recycling**.

**ALDFG**   An abbreviation for "abandoned lost or otherwise discarded fishing gear."

**Amine**   A family of chemical compounds whose molecules consist of an ammonia ($NH_3$) molecule, in which one hydrogen has been replaced by another species.

**Amorphous**   Without any crystalline structure.

**annealing**   A process for removing the stresses present in a material by repeated heating and cooling of the material.

**aramid**   A class of plastic resins with exceptional strength and thermal stability. The term is a shortened version of *aromatic polyamide*. Nylon is perhaps the best-known plastic made from aramid.

**biodegradable**   Organic material, such as plastics, that can be broken down by microorganisms into simpler, more stable compounds, also: see **bioplastics**.

**bioplastics**   A synonym for "biodegradable plastics."

**British thermal unit (Btu or BTU or btu)**   The quantity of heat needed to raise one pound of water by 1°F (specifically, from 58.5°F to 59.5°F).

**calendaring**   A process by which a heated plastic product is squeezed between heavy rollers into a thin sheet or film.

**catalysis**   The process by which a substance (the catalyst) is added to a chemical reaction to increase the rate at which that reaction occurs.

**chemical recycling**   The treatment of plastic wastes with chemical reactions that break the polymers of which they are made down into their constituent monomers, also known as **advanced recycling** or **tertiary recycling**.

**chemical species**   A general term used for many types of chemical substances, such as atoms, molecules, ions, or fragments of molecules.

**circular economy**   An economic system based on the elimination of wastes and the continual use of resources.

**clarity**   Absence of cloudiness in a plastic material or object.

**closed-loop recycling**   *See* **secondary recycling**.

**coefficient of friction**   A measure of the ease with which one object slides along the surface of another object.

**coefficient of thermal expansion**   A measure of the amount by which an object expands when heated by some given amount (usually, one degree Celsius).

**cold flow**   *See* **creep**.

**commingled containers**   A container that can be used for several different types of wastes intended for recycling. A typical such container might be allowed to hold aluminum and steel cans; glass bottles and jars; and plastic bottles, jars, jugs, and cups.

**copolymer**   A polymer made by the reaction between two different monomers.

**corrosion**   Changes that occur on the surface of an object because of exposure to heat or some chemical substance.

**crazing**   The appearance of fine cracks on or just beneath the surface of a plastic object.

**creep**   The extent to which a solid material undergoes deformation over a period of time, also called **cold flow**.

**cross-linking**   The process by which chemical bonds form between adjacent molecules in a substance. It usually results in the formation of a material that is resistant to melting and is insoluble in water.

**curbside recycling**   A recycling system in which plastics, glass, metal, and other recyclable products are placed in dedicated bins and placed alongside a road or street for pickup by a water management company.

**curing**   The process of changing a polymer's chemical, physical, electrical, or other properties by treatment with a chemical, application of heat or pressure, or some other action.

**density** The weight of some material or object per unit of volume, such as grams per cubic centimeter or pounds per cubic foot.

**destructor** The original name used for incinerating devices.

**drop-off recycling sites** Locations similar to open dumps or landfills to which recyclable materials can be delivered by consumers or businesses.

**elasticity** The tendency of an object to return to its original shape after being deformed by some physical process.

**elastomer** Any substance that is capable of returning to its original shape once it has been stretched to some significant amount.

**emulsion** The temporary mixture of two liquids that are insoluble in each other.

**end destination facility** Any factory, mill, or other facility where recyclable materials are converted into new products or raw materials.

**entanglement** The process by which an animal becomes trapped in some physical material or device, as, for example, when a turtle becomes trapped in a fishing net.

**epoxide** An organic compound containing a three-membered ring consisting of one oxygen atom and two carbon atoms.

**epoxy resins** A plastic material formed by the polymerization of epoxide structures.

**extender** A material that is added to a plastic to increase its bulk, reduce the cost of production, create or alter physical properties, or make some other desired change in the final product, also known as **filler**.

**extrusion** An important manufacturing process in which a heat-softened plastic is forced through an opening of some predetermined size and shape to form an object or material with that shape.

**fabricate** A general term that describes all of the steps involved in converting some raw material into some final product.

**fiber**    A thread-like material, such as silk, wood, or fiberglass.

**filler**    See **extender**.

**flash point**    The lowest temperature at which a material will produce sufficient vapor to allow a flame to develop.

**ghost fishing gear**    Fishing equipment, such as nets, pots, and traps, that has been intentionally discarded or accidentally lost in the oceans.

**gyre**    A large-scale system of currents produced by the action of surface winds over the oceans. There are five major gyres in the oceans: North and South Atlantic, North and South Pacific, and Indian. Gyres are a major site for the collection of plastic wastes.

**homopolymer**    A polymeric material made of a single kind of monomer.

**hydrocarbon**    A chemical compound consisting of the elements carbon and hydrogen only.

**hygroscopic**    The tendency to absorb moisture from the air.

**incineration**    The process of burning a substance, such as a waste product, for the purpose of reducing its volume and, perhaps, some of its hazardous characteristics. Heat produced by incineration is often used to produce electricity or for the heating of commercial or residential structures.

**inhibitor**    A substance that slows down the rate of a chemical reaction, such as the polymerization of a resin.

**injection molding**    A method for producing a plastic object of desired size and shape by forcing the molten plastic through a tube into a cavity of the shape and size desired.

**interaction**    The process by which an organisms comes into contact with some physical object without also becoming entangled in that object.

**landfill**    See **sanitary landfill**.

**light stability**    The ability of a plastic material to retain its color or other physical properties after exposure to sunlight or artificial lights.

**macroplastics**    Pieces of plastic greater than 25 millimeters in size.

**masterbatch**    A solid or liquid material added to a plastic to produce some desirable property, such as color, strength, resistance to bacterial attack, or oxidation.

**materials recovery facility**    A site at which commingled wastes are separated into individual components, such as paper, metal, and plastic, for recycling or other uses.

**mechanical recycling**    See **secondary recycling**.

**mermaid tears**    See **nurdle**.

**mesoplastics**    Small pieces of plastic with dimensions between 5 and 25 millimeters.

**microbead**    A small piece of plastic, generally less than 2 millimeters in size, made intentionally for use in a variety of cosmetic and personal care products.

**microplastics**    Very small pieces of plastic material that have entered the environment and broken down into pieces less than 5 millimeters long in any one direction. (Note that there is some disagreement as to the precise definition of this term.)

**molecular weight**    The weight of a molecule determined from the combined atomic weights of the atoms of which it is made. Thus, the molecular weight of ethylene ($C_2H_2$) is $2 \times 12$ (atomic weight of carbon) + $2 \times 1$ (atomic weight of hydrogen) = 26.

**monomer**    A chemical species from which a polymer can be formed.

**nanoplastics**    Unintentionally produced particles of size between 1 and 1000 nm that act like colloids in the environment. (Note that there is some disagreement as to the precise definition of this term.)

**NOAA**    Acronym for National Oceanic and Atmospheric Administration, the primary federal agency that oversees the health of the nation's waterways and water resources.

**nontoxic**   Not poisonous

**nurdle**   A pre-production microplastic pellet about the size of a pea. Sometimes also known as *mermaid tears.*

**olefin**   The common name for a family of organic compounds that contain one double bond per molecule. The two simplest olefins are ethylene (ethene) and propylene (propene).

**open dump**   The simplest form of waste disposal on land, consisting essentially of an open space on which wastes of any kind can be left, with no special treatment system, no method of collecting waste gases and solids, and access to anyone who wishes to use the facility.

**open-loop recycling**   See **primary recycling**.

**organic compound**   A chemical compound that contains one or more atoms of carbon. (A few important exceptions exist to this definition.)

**petrochemical**   A chemical substance obtained from the refining or processing of petroleum or natural gas.

**phenolic resins**   A plastic material that contains within its structure at least one phenol ($C_6H_5OH$) ring.

**photodegradable**   The ability of a material, such as a plastic, to decompose when exposed to light.

**plastic bag (or bottle) disposal law**   Legislation that prohibits landfill disposal of plastic bags, bottles, or other consumer products by generators or collectors.

**plasticizer**   A substance added to a plastic resin to increase its flexibility and to decrease brittleness.

**polymer**   A substance of large molecular weight produced by the combination of one or two simple compounds, known as monomers.

**postconsumer**   The name given to any type of waste that has been used for some purpose and then collected for treatment and reuse for some new and, perhaps, different purpose.

**postindustrial**   The name given to manufactured products or their by-products that fail to meet specifications, safety standards, or other criteria. They are normally collected and sold to specialized companies for conversion to other forms that do have some industrial use.

**primary microplastic**   Objects or materials specifically manufactured to have very small dimensions, such as the microplastics used in cosmetics.

**primary recycling**   The process of reusing a product, such as a plastic material, for the same purpose for which it was originally manufactured. Also known as **open-loop recycling**.

**quaternary recycling**   The use of plastic wastes for the generation of energy. Also known as **valorization**.

**recycling**   The process of treating a water material in some way as to make it used in new materials or new products.

**resin**   An organic substance, generally synthetic, which is used as a base material for the manufacture of some plastics.

**Resin Identification Code (RIC)**   A numerical coding system in which symbols and numbers are molded directly onto plastic bottles and containers to identify the resin from which they are made to indicate that they can be recycled in one specific way or another.

**sanitary landfill**   A waste disposal site that contains several features designed to prevent pollution of surrounding air and water, such as gas and leachate management, proper lining, compaction of wastes, daily and final covering, access control, and record keeping.

**saturated hydrocarbon**   A hydrocarbon in which all carbon atoms hold as many hydrogen atoms as possible.

**secondary microplastics**   Microplastics that are produced by the breakdown of larger plastic products, such as the changes that occur in a plastic object when exposed to the wave action of the oceans.

**secondary recycling**    The process of reusing a material, such as a plastic, for some purpose other than the one for which it was originally intended. Also known as **closed-loop recycling** or **mechanical recycling**.

**sludge**    Solid residue obtained from a wastewater or similar water treatment plant.

**slurry**    A watery mixture of some insoluble material.

**source reduction**    Any method designed to manufacture a product, such as a plastic object, in such a way as to result in smaller amounts of waste.

**source separation**    A type of recycling system in which different kinds of recyclable materials are separated by type at the source of generation.

**specific gravity**    The ratio of the density of a material compared to the density of water. Since the density of water is 1.00 grams per cubic centimeter, the specific gravity of solids and liquids is numerically the same, with the exception that the latter has no units of measure attached to it.

**stabilizer**    A material added to a plastic to reduce its tendency to undergo change, such as loss of color.

**surfactant**    A material that reduces the force of repulsion between two objects or materials, such as the presence of a detergent (the surfactant) in a mixture of oil and water, otherwise insoluble in each other.

**sustainability**    As defined by the United Nations World Commission on Environment and Development, sustainability means "development that meets the needs of the present without compromising the ability of future generations to meet their own needs."

**tensile strength**    The resistance of a material to tearing or breaking apart when exposed to external forces.

**tertiary recycling**    See **chemical recycling**.

**thermal conductivity**  The ability of and extent to which a material will conduct heat.

**thermoforming**  The process of heating a thermoplastic material to a working temperature and then forming it into a finished shape by means of heat or pressure.

**thermoplastic**  The name of a process or a material that can be repeatedly melted and re-formed.

**thermoset**  The name of a process or a material that, once formed, cannot be reheated and re-formed.

**unsaturated hydrocarbon**  A hydrocarbon in which one or more carbon atoms contain fewer hydrogen atoms than possible.

**urea-formaldehyde plastic**  A polymer made in the reaction between urea, $CH_4N_2O$, and formaldehyde, $CH_2O$,

**UV stabilizer**  Any material added to a plastic to reduce like-lihood of damage resulting from exposure to ultraviolet (UV) radiation.

**valorization**  See **quaternary recycling**.

**virgin plastic**  Any plastic produced by a polymerization reaction that has not undergone any changes or modifications and contains no recycled plastic.

**viscosity**  The resistance of a liquid to flow.

**waste diversion**  Any practice in which waste materials are delivered to some operation other than incineration or landfill for disposal.

**waste-to-energy plant**  A facility at which wastes are burned to produce heat or electrical energy for use at some other location.

**zero waste**  The concept of, as well as the name of organizations interested in, the reduction of solid waste generation to zero, or at least as close to that goal as possible. With organizations, the term is capitalized: Zero Waste.

Note: Page numbers followed by *t* indicate tables.

## About the Author

**David E. Newton** holds an associate's degree in science from Grand Rapids (Michigan) Junior College, a BA in chemistry (with high distinction), an MA in education from the University of Michigan, and an EdD in science education from Harvard University. He is the author of more than 400 textbooks, encyclopedias, resource books, research manuals, laboratory manuals, trade books, and other educational materials.

He taught mathematics, chemistry, and physical science in Grand Rapids, Michigan, for 13 years; was professor of chemistry and physics at Salem State College in Massachusetts for 15 years; and was adjunct professor in the College of Professional Studies at the University of San Francisco for 10 years.

Some of the author's previous books for ABC-CLIO include *Eating Disorders in America* (2019), *Natural Disasters* (2019), *Vegetarianism and Veganism* (2019), *Gender Inequality* (2019), *Birth Control* (2019), *The Climate Change Debate* (2020), *World Oceans* (2020), *GMO Food* (2021), and *Hate Groups* (2021). His other books include *Physics: Oryx Frontiers of Science Series* (2000), *Sick!* (4 vols., 2000), *Science, Technology, and Society: The Impact of Science in the 19th Century* (2 vols., 2001), *Encyclopedia of Fire* (2002), *Molecular Nanotechnology: Oryx Frontiers of Science Series* (2002), *Encyclopedia of Water* (2003), *Encyclopedia of Air* (2004), *The New Chemistry* (6 vols., 2007), *Nuclear Power* (2005), *Stem Cell Research* (2006), *Latinos in the Sciences, Math, and Professions* (2007), and *DNA Evidence and Forensic Science* (2008). He has also been an updating and consulting editor on a number of books and reference works, including *Chemical Compounds* (2005), *Chemical Elements* (2006), *Encyclopedia of Endangered Species* (2006), *World of Mathematics* (2006), *World of Chemistry* (2006), *World of Health* (2006), *UXL Encyclopedia of Science* (2007), *Alternative Medicine* (2008), *Grzimek's Animal Life Encyclopedia* (2009), *Community Health* (2009), *Genetic Medicine* (2009), *The Gale Encyclopedia of Medicine* (2010–2011), *The Gale Encyclopedia of Alternative Medicine* (2013), *Discoveries in Modern Science: Exploration, Invention, and Technology* (2013–2014), and *Science in Context* (2013–2014).